Reliability by Design

Reliability by Design
CAE Techniques for Electronic Components and Systems

A.C. Brombacher

Philips Consumer Electronics
The Netherlands

JOHN WILEY & SONS
Chichester • New York • Brisbane • Toronto • Singapore

Other Wiley Editorial Offices

John Wiley & Sons, Inc., 605 Third Avenue,
New York, NY 10158-0012, USA

Jacaranda Wiley Ltd, G.P.O. Box 859, Brisbane,
Queensland 4001, Australia

John Wiley & Sons (Canada) Ltd, 22 Worcester Road,
Rexdale, Ontario M9W 1L1, Canada

John Wiley & Sons (SEA) Pte Ltd, 37 Jalan Pemimpin 05-04,
Block B, Union Industrial Building, Singapore 2057

**A catalogue record is available
from the British Library**

ISBN 0 471 93193 4

Produced from camera ready copy supplied by the author
Printed in Great Britain by Courier International, East Kilbride

To Ineke

Contents

Preface   **xiii**

Acknowledgements   **xvii**

1. Introduction   **1**
  1.1. Design optimization   2
  1.2. Reliability   5
  1.3. Reliability analysis used in the design process   7

**2. The usability of existing reliability prediction methods
     for reliability optimization**   **9**
  2.1. Introduction   9
  2.2. Existing reliability prediction methods   9
      2.2.1. Statistical evaluation methods used in the development of
              reliability prediction models   12
      2.2.2. United States Department of Defence MIL-HDBK-217   15
      2.2.3. British Telecom HRD-4   18
      2.2.4. Other reliability models for integrated circuits   20
          2.2.4.1. The Philips failure rate model for integrated circuits   20
          2.2.4.2. Kasouf & Mercurio model for integrated circuits   20
          2.2.4.3. The CNET failure rate model for integrated circuits   21
          2.2.4.4. The extended MIL-HDBK-217E model for integrated circuits
                    (Pantic model)   22
      2.2.5. Summary of standard failure rate handbooks   22
  2.3. Reliability calculations using standard reliability prediction handbooks   24
      2.3.1. Reliability prediction of test circuits compared to practical
              reliability figures   24
          2.3.1.1. Reliability prediction test circuits A and B   25
          2.3.1.2. Reliability prediction test circuit C   27
      2.3.2. Summary of practical reliability predictions   29
  2.4. Reliability optimization according to standard reliability prediction
        handbooks   29
      2.4.1. Parameters in existing reliability models usable for reliability
              optimization   30
      2.4.2. Reliability optimizations derived from standard reliability
              prediction models   33
      2.4.3. Summary of reliability optimization using existing reliability
              prediction models   34

3. Failure prediction using stressor/susceptibility interaction   **39**
3.1. Introduction   39
3.2. Part failure mechanisms   40
3.3. Stressors   41
    3.3.1. Stressors defined as stochastic functions   42
    3.3.2. Stressor probability density function; single circuit, single mode   42
    3.3.3. Stressor probability density function; single circuit, multiple modes 45
    3.3.4. Stressor probability density function; multiple circuits, multiple
            modes   47
    3.3.5. Multi-variable stressor probability density functions
            (practical example)   49
    3.3.6. Examples of practical stressors   51
        3.3.6.1. Electrical stressors   51
        3.3.6.2. Thermal stressors   53
        3.3.6.3. Mechanical stressors   53
3.4. Susceptibility for (combinations of-) stressors   53
    3.4.1. One variable catastrophic susceptibility model   54
    3.4.2. Multi-variable catastrophic susceptibility models   55
    3.4.3. Gradual susceptibility models   56
    3.4.4. Constantly degrading susceptibility models   58
    3.4.5. Susceptibility models for large series components   59
    3.4.6. Weak sub-populations   60
3.5. Failure probability and reliability   60
    3.5.1. Failure probability for single failure mechanisms   61
    3.5.2. Component failure probability for multiple failure mechanisms   65
    3.5.3. Components with identical constant susceptibility   69
    3.5.4. Components with different but constant susceptibility   73
    3.5.5. Weak sub-populations   74
    3.5.6. Degradation effects   75
    3.5.7. Gradual failure mechanisms, cumulative effects   76
    3.5.8. Combined effects   77
3.6. Failure probabilities in terms of design parameters   77
3.7. Summary of failure prediction   81

4. Deriving susceptibility models from failure mechanisms   **83**
4.1. Introduction   83
4.2. Failure mechanisms in electronic components   85
4.3. Electrical overstress failure mechanisms   85
    4.3.1. Thermal considerations   85
    4.3.2. Current breakdown (Hot-spot melting)   89
    4.3.3. Power breakdown (thermal cracks)   91
    4.3.4. High-voltage breakdown   93
        4.3.4.1. Impact ionization   93
        4.3.4.2. Avalanche and Zener breakdown   94
        4.3.4.3. Electron-trap ionization   95
4.4. Long term failure mechanisms   96
    4.4.1. Corrosion   97
    4.4.2. Electromigration   98
    4.4.3. Secondary diffusion   99

4.5.   Additional failure mechanisms for bipolar semiconductors 100
    4.5.1.   Pulse power effects 100
    4.5.2.   Second breakdown 103
        4.5.2.1.   Geometrical transistor aspects related to breakdown effects 106
        4.5.2.2.   Forward-bias second breakdown 109
        4.5.2.3.   Reverse-bias second breakdown 111
    4.5.3.   Summary of the discussed failure mechanisms 114
4.6.   Susceptibility models for practical components 116
    4.6.1.   Diode X (schottky diode) 116
        4.6.1.1.   Pulse power effects 117
        4.6.1.2.   Current breakdown 118
        4.6.1.3.   Avalanche breakdown 119
        4.6.1.4.   Power breakdown 119
    4.6.2.   High voltage transistor Y 120
        4.6.2.1.   Current breakdown 121
        4.6.2.2.   Power breakdown & secondary diffusion 121
        4.6.2.3.   Avalanche breakdown 121
        4.6.2.4.   Forward bias second breakdown 123
        4.6.2.5.   Reverse bias second breakdown 123
    4.6.3.   Integrated circuit Z (motor driver IC) 124
        4.6.3.1.   Power dissipation and secondary diffusion 130
        4.6.3.2.   Current breakdown and electromigration 130
        4.6.3.3.   Avalanche breakdown 130
        4.6.3.4.   Flyback diodes D1 & D2, pulse power effects 130
        4.6.3.5.   Switching transistors T1 & T2, second breakdown 131
4.7.   Summary of susceptibility models 133

5.   Stressor sets for practical circuits **135**
5.1.   Introduction 135
5.2.   Acquiring stressor sets 139
5.3.   Deriving stressor sets from computer simulation results 140
    5.3.1.   Requirements for simulation software 141
    5.3.2.   Requirements on functional component models 143
    5.3.3.   Parameters required for simulation models 144
    5.3.4.   Requirements for component tolerance models 145
    5.3.5.   Summary of the demands on circuit simulation for stressor/
        susceptibility analysis 147
5.4.   Deriving stressor sets from practical measurements 148
    5.4.1.   Requirements for measurement hardware 148
    5.4.2.   Measurement of individual stressor-sets 150
    5.4.3.   Measurements of mean stressor-sets 151
        5.4.3.1.   The use of pre-selected circuits 152
        5.4.3.2.   Relating failures in feedback circuits to mean stressor-sets 153
5.5.   Practical stressor/susceptibility interactions 156
    5.5.1.   Diode X, circuit A 157
        5.5.1.1.   Individual stressor set 159
        5.5.1.2.   Mean stressor-set 160
        5.5.1.3.   Stressor/susceptibility interaction 162
    5.5.2.   Transistor Y in circuits A and B 164
        5.5.2.1.   Individual stressor set 165

	5.5.2.2.	Tolerance effects of transistor Y	169
	5.5.2.3.	Tolerance influence of other components	173
	5.5.2.4.	Relation to time-failure probability	173
5.6.	Summary of practical stressor/susceptibility interaction		174

6. Reliability optimization using stressor/susceptibility models   **175**

6.1.	Introduction		175
6.2.	Deriving stressor sets from circuit simulation		178
	6.2.1.	Worst-case analysis	178
	6.2.2.	Parameter regionalization	179
	6.2.3.	Monte Carlo Analysis	181
	6.2.4.	Pass/fail diagrams	182
6.3.	Reliability optimization using the centre of gravity method		185
	6.3.1.	Long-term stressor and susceptibility models in reliability optimization	189
	6.3.2.	Suggestions for enhancement of the CoG method	190
	6.3.3.	Tolerance models required for reliability optimization	194
6.4.	Practical example		195
	6.4.1.	Parameter tolerances	196
	6.4.2.	Optimization using the Centre of Gravity method	197

7. Conclusions   **201**

7.1.	Impossibility to use existing reliability prediction methods	201
7.2.	New method: stressor/susceptibility interaction	202
7.3.	Practical use of stressor/susceptibility models	203
7.4.	Development of susceptibility models	203
7.5.	Stressor sets	204
7.6.	Reliability optimization	204
7.7.	Practical use of stressor/susceptibility analysis in industry	205
7.8.	Recommendations for further research	206

A. Short explanation of the test circuits used   **207**

A.1.	Introduction		207
A.2.	Practical circuits		207
A.3.	Reliability predictions for circuits used in consumer electronics		208
A.4.	Circuit A		209
	A.4.1.	Functional structure circuit A	210
	A.4.2.	Components used (reliability aspects)	213
	A.4.3.	Reliability prediction using existing prediction methods	215
	A.4.4.	Practical failure data	216
A.5.	Circuit B		218
	A.5.1.	Functional structure circuit B	219
	A.5.2.	Components used (reliability aspects)	219
	A.5.3.	Reliability prediction using existing prediction methods	222
	A.5.4.	Practical failure data	223
A.6.	Circuit C		224
	A.6.1.	Function	224
		A.6.1.1. The motor drive IC	225

A.6.1.2.   Function of the motordrive IC in circuit C 227
A.6.2.   Components used (reliability aspects) 227
A.6.3.   Reliability prediction using existing prediction methods 228
A.6.3.1.   The MIL-HDBK-217 motor model 229
A.6.3.2.   Reliability figures used components 230

B.   Failure mechanisms in simple components 233
B.1.   Resistive components 233
B.1.1.   General failure mechanisms 233
B.1.1.1.   Power overstress 233
B.1.1.2.   Pulse power effects 234
B.1.1.3.   Effects of inhomogeneities on power dissipation 237
B.1.1.4.   Voltage overstress 237
B.1.2.   Practical components 238
B.1.2.1.   Wire wound resistors 238
B.1.2.2.   Film resistors 239
B.2.   Capacitive components 241
B.2.1.   General failure mechanisms 242
B.2.1.1.   High-voltage breakdown 242
B.2.1.2.   Effects of inhomogeneous structures on voltage breakdown 243
B.2.1.3.   Power breakdown 243
B.2.2.   Practical components 243
B.2.2.1.   Ceramic/plastic capacitors 243
B.2.2.2.   Electrolytic capacitors 244
B.3.   Inductive components 247
B.3.1.   Single air coils 248
B.3.2.   Multiple air coils 249
B.3.3.   Inductive devices using magnetic cores (coils, transformers) 250
B.4.   Conclusions 252

C.   Tolerance models and examples 253
C.1.   Introduction 253
C.2.   Example 1: Foil transformer (circuit B) 254
C.3.   Example 2: The optocoupler (circuit B) 256
C.4.   Example 3: High voltage transistor 260
C.5.   Conclusions 261

D.   Bibliography 263

E.   Terms used 267
E.1.   General terms 267
E.2.   Stressor/susceptibility 267
E.3.   Traditional reliability prediction models 268
E.4.   Electrical variables 269

Index 271

Preface

This book describes a method for the use of reliability analysis in the design process of electronic circuits and systems. As this method was developed especially for high-volume consumer electronics this book will emphasize especially of this category of electronic circuits. It is nevertheless possible to use this method also for other types of electronic circuits. The presented method was tested on three practical circuits: two switch mode power supplies and one motordrive circuit. All circuits are used in the current generation of Philips video cassette recorders.

Chapter 1 describes the current position of reliability in the production process of consumer electronics. It emphasizes the need for reliability analysis and optimization as part of the design process and defines the demands for such an integrated reliability analysis and optimization.

The second chapter of this book investigates the possibilities of using existing reliability analysis methods for this purpose. First the backgrounds of existing reliability analysis methods are explained with emphasis on the mechanisms used in the existing prediction models. In the second part of this chapter a reliability prediction is performed for the three circuits mentioned above. Results of the predictions are compared and then compared with *practical data*. The result of this analysis is that existing reliability analysis methods are not useable for reliability optimization purposes. First of all it is not possible to derive one common denominator from the existing models; second the differences between the practical figures and the predicted figures are too large for useful application. Finally the number of influence factors used in the existing models is insufficient for a detailed optimization.

In the third chapter a theoretical method is developed describing reliability analysis based on what is called *stressor/susceptibility* interaction:

- Stressors are defined as influence factors having an influence on the failure probability of a component

- Susceptibility is defined as the probability that a component will fail under a certain combination of stressors

A more detailed definition of these terms is given in Chapter 3. The advantage of this method is the possibility of taking tolerance and drift effects of functional component parameters into account.

The development of susceptibility models for practical components is presented in Chapter 4. The first part of this chapter explains some well known failure mechanisms in terms of stressors while the second part of this chapter uses these failure mechanisms in the development of examples of stressor/suscepti-bility models for practical components. Three sample components are used: a medium power Schottky diode, a high-voltage bipolar transistor and an integrated circuit. The three components where selected because of either the differences in reliability models (presented in Chapter 2) or because of considerable differen-ces between predicted and practical reliability figures. Unfortunately at the moment of writing of this book insufficient samples of the integrated circuit were available for the development of a complete susceptibility model.

The fifth chapter describes the use of stressor/susceptibility models in the analysis of practical circuits.An important part of this chapter is the derivation of stressor sets for certain components in these circuits. Important aspect of this chapter is the fact that for many practical components used in modern electronics no usable functional tolerance model is available. Therefore, before going into detail of stressors, this chapter describes the requirements for tolerance models of electronic components. (Practical tolerance models are presented in detail in Appendix C.) An important conclusion from this chapter is: It is possible to use stressor/susceptibility models in the analysis of component failure causes within a circuit. Remaining problem, however, is the fact that the time- dependency remains a difficulty to be solved. Analysis of practical circuit failures proves that for the tested circuits a vast majority of the reliability problems can be explained using stressor/susceptibility interaction.

The sixth chapter of this book presents a method to use stressor/susceptibility interaction in circuit optimization. As stressors and susceptibility of components, in contrast with the traditional reliability models, can be expressed in terms of designable parameters, it is possible to optimize circuit parameters in order to achieve maximum reliability. The method used for this reliability optimization is closely related to the optimization of component tolerances in order to achieve optimum yield. The suggested optimization method is illustrated using a simple practical example.

Chapter 7 finally presents conclusions of this research project and gives recom-mendations for further research.

Appendix A presents in detail reliability predictions of the used sample circuits in accordance with the traditional reliability prediction handbooks BT-HRD4 of

British Telecom and MIL-HDBK-217E of the United States Department of Defence.

Appendix B gives, in the form of a sort of "cookbook recipes", guidelines for the development of susceptibility models for components other than those discussed in Chapter 4.

As, during this research project, it turned out that useable tolerance models for many components are not available, Appendix C presents requirements for the development of tolerance models in general and discusses practical tolerance models of some components analysed during this project. The chapter focuses on tolerance models of a bipolar transistor, tolerance models of an optocoupler and the tolerance model of a foil transformer.

Acknowledgements

Herewith I would like to express my gratitude for the important support of many people. Without their help it would have been impossible to complete this project. First of all I want to thank Prof. Dr.-Ing. O.E. Herrmann of Twente University for his coordination and support and, especially, for giving the finishing touch to my book. He literally dotted my i's and crossed my t's. I would like to thank Prof. Dr J.F. Verwey for the interesting discussions on the subject and the support of the Philips NatLab, especially on the subject of failure verification.

I would especially like to thank Mr J. van't Loo of the Philips M.R. Laboratory. His enthusiastic support was essential for the entire project.

I appreciate the motivating discussions with Mr A. Harris and his support from the side of the Philips management. An important part of this project was carried out at the Philips Magnetic Recording laboratory (former VIDEQ lab.) Therefore I would like to thank Ir. H. Post, Ir. Th. Vlek and Ir. Th. Martens as department heads of this laboratory for allowing me to use the facilities of the MR laboratory and for their important support, especially on the subject of failure verification and reliability improvements. I would like to thank Ir. P. van Leeuwen for his support and the valuable discussions on the subject of what is called in this book circuit C and IC Z. I appreciate the assistance of Mr G. van Schaik of Philips Components Investigation and Reliability Group (CIRG) and other people of Philips CIRG on the subject of component failures.

Dipl.-Ing. H. Lhotak and Ing. E. Severa of Philips VIW Vienna supported this project especially on the subject of failure verification and reliability improvements. I do hope that the results obtained in this project are of use in Vienna.

I would like to thank Mr M. Bennion, Dr J. Humphreys and Ir. T. van der Wouw for their important contributions on the subject of failure mechanisms in bipolar components. Their knowledge and support proved to be of great value. I would like to thank Mr A. Marinus of the Philips VDP laboratory for the useful discussions on what is called in this book circuit A and circuit B.

The method for reliability optimization, presented in Chapter 6 of this book, is closely related to the work performed by Prof. R. Spence of Imperial College in

London. I want to thank Prof. Spence for the very interesting and useful discussions on tolerance design. An important part of the calculations presented in Chapter 6 of this book was performed using a specially adapted form of the MINNIE software package. I would like to thank Mr P. Jennings of ISL in London for the software support.

I wish to thank my mother, Mrs J.C.A. Brombacher-van Asch, for her assistance with the English language and for corrections of the text of this book.

Many students have cooperated in this project. I wish to thank them for their often very enthusiastic cooperation. I wish to thank Ing. B. Schouten and Ing. J.A.M. Molkenboer for their contributions, especially in the earlier phases of this project.

Finally I would like to thank the people who worked with me on this project, often for a period of years. This book is also based on their work. Rik de Boer, Peter Fennema, Maarten Gommers, Roelof Vos and Jacques Walinga, I thank you all for your important participation in this project and for the commitment you showed.

1
Introduction

In common language there is a wide range of activities associated with the word 'design'.The word 'design'is used not only to cover the creative process of making a sculpture but also to describe the technical process of making a satellite. Although there is a tremendous difference in these activities they have some aspects in common.

Generally speaking every design process starts with a number of *demands*. The aim of the design process is to combine these demands into a concept for a product.

A major problem in designing is the fact that often the various demands will conflict. One of the best known examples in this respect is the conflict between the cost and the performance. Therefore in practice designing is finding a compromise between a number of, often contradictory, demands. This, and the fact that often during the design process additional demands are found, will result in a design process consisting of many changes. For every step of the design process all these different demands should be considered, each with its own weight factor, in order to find an optimal design. In other words, designing will be the solving of an optimization problem with many constrained variables.

Although there is a wide variety of demand functions influencing the resulting product it is, at least for electronic circuits and systems, important that the following criteria are fulfilled.

— The system should fulfil its intended purpose. (It should work.)

— The system should result in minimum cost. (It should be cheap.)

— The system should fulfil this purpose for a certain period of time. (It should remain working.)

For systems produced in large quantities the previous statements will change slightly:

— A vast majority of the produced systems should fulfil their intended purpose.

— The design and production of the system should result in minimum overall cost.

— A vast majority of the produced systems should fulfil their purposes for a certain period of time.

In other words, it is possible to use three dominant demand areas: *Cost*, *Performance* and *Reliability*. See figure 1.1. Of course the way in which these individual demands influence the design process depends strongly on the kind of product. The design of medical electronics will have more emphasis on the reliability of a product while, for example, high- volume consumer electronics will have more emphasis on cost minimization.

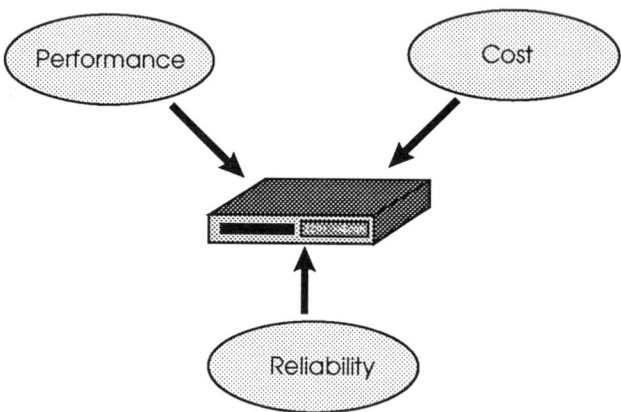

Figure 1.1: Dominant demand areas

Traditionally functional system design (performance) has always been one of the most important topics in system design. Cost minimization is also a well known topic in the design process. Reliability analysis, on the other hand, is seldom used during the actual system design.

1.1. Design optimization

One of the most characteristic aspects of the development process of a system is the fact that only in a few cases a design is completed without numerous, often tedious, changes. These changes are caused by the fact that all development activities consist of two phases: synthesis and verification. In those cases where a verification step shows deficiencies in a given design it will be necessary to repeat one or more synthesis steps. As the changes are initiated by numerous causes, they will be introduced in the development process somewhere between preliminary system design and system use. See figure 1.2.

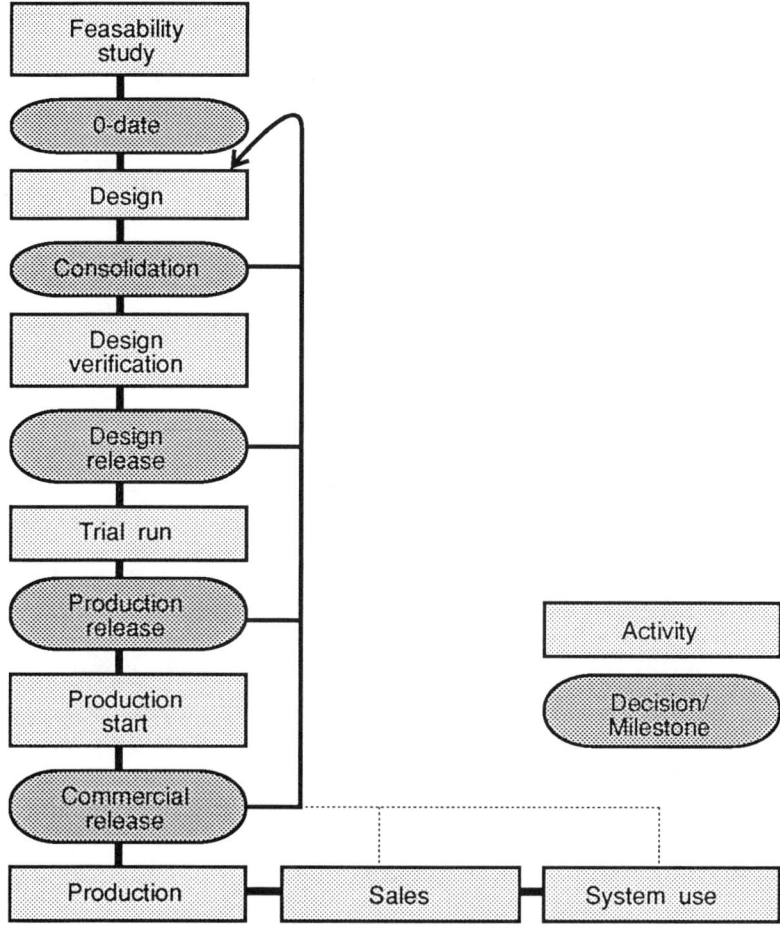

Figure 1.2: From planning to use

There is a considerable difference in which phase of the development process the design changes are introduced. In general, modifications are introduced in the first phases of the development cycle. Changes introduced later do occur but should be avoided due to the great effort required to modify an almost completed design. This is due to the large number of parameters involved. The problem will often be that there is quite a difference in one single, working, prototype in a laboratory environment and a large batch of systems working according to their specifications for a stated period of time under certain environmental conditions.

Generally speaking there are two methods in use to predict the functional behaviour of a batch of systems:

— Prediction using simulation models and techniques.

— Intensive tests of prototypes.

An advantage of the first method is the possibility of performing system optimizations during an early design phase. If there are prediction models available expressing wanted and unwanted effects in terms of designable parameters, it is possible that bottlenecks are removed and performance is improved by the designers themselves as an integrated part of the design process. The term *designable* is used to indicate those parameters where a designer has the freedom to change parameter values (or design structures) in order to achieve a certain effect. An advantage of this comprehensive optimization is the fact that multiple constraints can be analyzed in one optimization step. This is called *on-line* optimization [Tri88].

An advantage of the second method is the fact that, generally speaking, many design weaknesses are detected. A disadvantage is that this method is rather time consuming and tedious. It requires quite a lot of effort to modify an (often) almost completed design. Therefore the main aim of optimization will now be to remove or suppress the unwanted effect. Generally speaking this tends to result in the addition of constructions to suppress the deficiencies and leaving the rest of the design structure as much as possible intact. A second consequence will be that, due to the mature state of the design, a lot of interaction between different groups of people is required for example, designers, quality assurance teams, production teams, logistic departments, component suppliers; depending on the state of the project where changes are introduced. This is a second reason to minimize the number of changes required to remove or suppress the unwanted effects. As a result solutions will be *added to* and will not be *integrated in* a design. This is called *off- line* optimization.

> *There is a considerable difference in the "quality" of design*   (1.1)
> *improvements **integrated in** a design compared to the (often*
> *ad-hoc) design improvements **added to** a design.*

Predictions are therefore used to improve the efficiency of the design process. This has resulted in the introduction of a large number of, especially computer aided, simulation tools in the last two decades. A problem of the use of prediction models is the fact that, generally speaking, not all possible problems are predicted. Therefore functional tests are used for control and verification purposes to find problems not predicted by the first method. These functional tests vary between the testing of one single circuit using a simple breadboard on one hand and user-simulation tests of several hundreds of systems on the other. These user-simulation tests emulate the conditions of actual system use for a large batch of systems.

1.2. Reliability

Traditionally reliability calculations and predictions are used in the design of systems where reliability is very important, for example military systems and medical equipment. For those systems unexpected failures are simply not allowed and should therefore be prevented, even if this results in higher costs or additional (development-) time required to prevent reliability problems.

One of the oldest and safest methods to check a system's reliability is to take a completed or almost completed system (or, for statistical reasons, take a complete (prototype) batch of such a system) and to test this system for a certain period of time. Therefore many design teams in many branches of industry have reliability control groups checking the reliability of the completed of (almost) completed designs. A disadvantage of this method is that the first complete reliability analysis of a design is performed on the completed (or almost completed) system. Therefore it will be tedious and expensive to introduce changes using this optimization method.

Other branches within the electronics industry have another priority for the reliability of their systems. For consumer electronics, for example, reliability is important but if additional costs, introduced by reliability measures will make a system less competitive there is a serious problem. Another problem is the fact that often a tremendous time pressure exists to have a system commercially available before the possible competitors. Therefore it will, generally speaking, hardly be possible to use the safe but tedious reliability assurance methods described above.

A first step to simplify this operation is to use the so-called "method of reliability prediction". This method was introduced during the late 1960s with the now historical "Military standarization handbook Reliability Prediction of Electronic Equipment MIL-HDBK-217A" [MIL65] and was first used in the design of military and aerospace equipment. This reliability prediction method gives a relation between environmental factors and reliability on the level of electronic components. Using this prediction method it becomes possible to predict the reliability of a circuit or system in the early phases of the design process. The formulas used are often based on a combination of practical experience and physical knowledge.

This reliability prediction method is used during design verification to analyze the reliability of the proposed circuit. The purpose of reliability analysis is to verify whether the predicted reliability of the circuits meets with the reliability demands. If the predicted reliability is in accordance with the demands the design is accepted. In other cases the design is rejected. The position of reliability analysis in the design process remained the same. As the job of predicting reliability using reliability prediction models is quite a tedious one reliability prediction is usually carried out by a group of specialists. In this situation the reliability of a design is predicted by a separate quality or reliability control group and reliability problems are returned to the functional design group in the form of some *added demands*. See also the previous paragraph. (See figure 1.3)

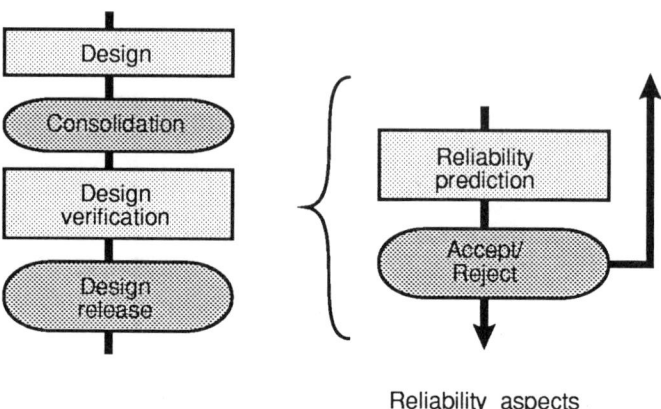

Reliability aspects

Figure 1.3: Reliability analysis causing extra demands

The advantage of this method is the possibility of introducing changes due to reliability demands shortly after the design synthesis phase. Due to the smaller amount of effort required to introduce changes reliability prediction will result in a considerable improvement compared to reliability testing. At this moment reliability prediction methods are used by both the traditional "high-rel" design teams and the design teams working at "commercial/consumer" products. Although the method of the use of reliability predictions is faster than the previously mentioned test method the work is still tedious and optimization will still be a matter of adding demands to a largely completed design. Another problem of the use of prediction models lies in the accuracy of the used prediction models. Although the reliability prediction techniques have increased in accuracy during recent years it is not (yet) possible to predict every possible reliability problem. There will always remain a number of "unexpected" failures. To overcome this problem often a combination of reliability prediction and reliability testing is used. Reliability prediction will be used to find a majority of the added reliability demands while tests will be used to find other, "unexpected", reliability problems.

The problem with this kind of added demands is the fact that, generally speaking, added demands will lead to (ad hoc) additions to a design. The question is: in how far will these additions result in a better design. The previous paragraphs defined designing as a constrained optimization process. The practice on the other hand shows that designing, especially where reliability parameters are involved, is in fact a sequential optimization process; an (almost) completed design is checked using reliability prediction techniques and either accepted or rejected. See figures 1.3 and 1.4.

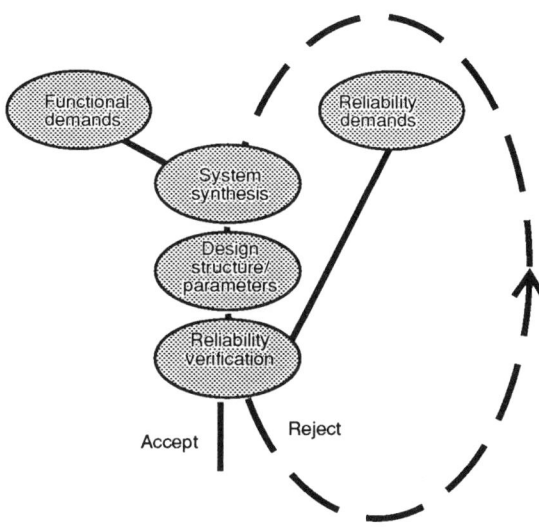

Figure 1.4: Current position of reliability analysis

1.3. **Reliability analysis used in the design process**

A better method would be to express reliability demands in terms of functional parameters. In this case the aim of reliability analysis is not to accept or reject a given design but to present reliability problems in terms of designable parameters. In this way it is possible to use one functional design phase of which the former reliability analysis is an integrated part. The advantage of this method is the possibility of integrating reliability optimizations in the design of a circuit or system. To use reliability analysis as part of the functional system synthesis results in some important demands.

— The reliability analysis method should be able to cover a majority of the practical reliability problems.

— It should be possible to relate reliability parameters directly to parameters used in the functional design synthesis.

An additional demand is related to the fact that, generally speaking, functional designers do not tend to have a reliability analysis background. Therefore it would be useful if the reliability analysis method would be transparent to the user. An important advantage in this respect are methods where possibilities exist to embed the method within existing functional computer aided design software.

The main purpose of this book is to find or develop reliability analysis methods useable in the comprehensive system synthesis described in this chapter. One of the important aspects of this book is the practical usability of the presented method. Therefore all procedures and models discussed in this book were tested on three practical circuits. The three test circuits will act as a guideline throughout this book.

As, at this moment, many well established methods for reliability prediction are available this book will first concentrate on these traditional methods, their backgrounds and the usability of these methods in system synthesis. One of the conclusions of this book is that these traditional methods are not usable in design optimization. Therefore the next part of this book will present a new reliability analysis method using so- called *stressor/susceptibility* interaction. The rest of this book will concentrate on:

— the development of practical models for this new method,

— comparing the prediction results obtained using this method to reliability problems occurring in practical situations,

— the possibilities of using the presented method in a comprehensive design synthesis technique.

2

The Usability of Existing Reliability Prediction Methods for Reliability Optimization

2.1. Introduction

As discussed in the previous chapter designing a circuit with "built-in" reliability will require reliability optimization in the early phases of the design process. One of the most important problems in this respect is the difficulty of estimating or analyzing a circuits reliability at a moment when no detailed field failure data are available. It is even more difficult to find the main influence factors required for optimization of the reliability of such a circuit at such an early moment of a circuit's life-cycle.

As, at the moment, many handbooks and standard methods for reliability prediction are available, this chapter will discuss the (im)possibility of using one of the existing methods for reliability prediction in the process of reliability optimization. To test the usability of standard reliability prediction methods, some examples will be presented. In these examples a standard reliability prediction will be made for a practical circuit. The results of these predictions will be used to derive guidelines for reliability optimization. Both the reliability prediction and the derived reliability optimization results will be compared to the results of actual field failures. The last part of this chapter will explain the fundamental differences between predicted and actual reliability figures and will illustrate the impossibility of using the current reliability prediction handbooks in the process of reliability optimization. Finally, detailed requirements for the development of new reliability optimization models will be given.

2.2. Existing reliability prediction methods

As, during the last decades, electronic circuits started to play an important role in many parts of every days life, it became more and more important to find out how reliable those electronic circuits were. Customers demanded in many cases not only a functional specification of a circuit but also a specification of the reliability of a circuit. As mentioned in the previous chapter it will be difficult to estimate the reliability of a circuit which does not yet exist. Therefore it became necessary to develop models describing the reliability of electronic circuits.

Physical defects on part level are the lowest level in the hierarchy of failures [Jon89]. Therefore most reliability prediction techniques are based on the prediction of part failures [Sho68] [Sie82]. The used part failure prediction models are based on statistical evaluation of actual part failures. The required part failure data is obtained from two main data sources:

— Field failures

— Failures obtained using reliability tests

Field failures are components within circuits, failing after a certain period of time within circuits under "normal" circumstances. If sufficient data are also available it will be possible to give a relation between part failure probability and time. This relation is only valid for components used under conditions similar to the original circumstances. If data from components failing under different circumstances are available it will be possible to develop models describing the influence of components being used under varying circumstances. The development of part failure prediction models using field failure feedback is, however, a very tedious and inaccurate job. Tedious because of the number of part failures required to obtain a statistical acceptable model will require the observation of large numbers of circuits over a large period of time. Reporting not the exact circumstances under which the failure occurred may lead to inaccuracies in the resulting model. A third problem of the development of reliability models using failure feedback is the often ironical situation that the resulting models have a maximum of accuracy at the moment the component is no longer used.

To prevent these problems, methods were needed to develop reliability models faster and with greater accuracy. Supposing that reliability problems in electronic components are due to certain chemical degradation mechanisms[*] it becomes possible to relate the temperature dependence of the reliability problem to the temperature dependence of the chemical process of this degradation mechanism. The reaction rate of such a chemical process is described by means of the Arrhenius Law:

$$K = A\, e^{\frac{-E_a}{kT}} \qquad (2.1)$$

K: Temperature reaction rate

A: Constant

E_a: Activation energy of the process

---

[*] Chapter 4 will show that this often used assumption is not always correct; there are also failure mechanisms governed by other processes than chemical degradation

k: Boltzmann constant

T: Absolute temperature

A consequence of the assumption that chemical reactions have a dominant influence on the failure behaviour of electronic components is the possibility of accelerating the tests required for the development of reliability models. Increasing the temperature will increase the reaction rate of the chemical degradation process. Due to the exponential form of this equation it is possible to achieve large acceleration factors by means of increasing the temperature thus shortening the observation time required to obtain an accurate reliability prediction model.

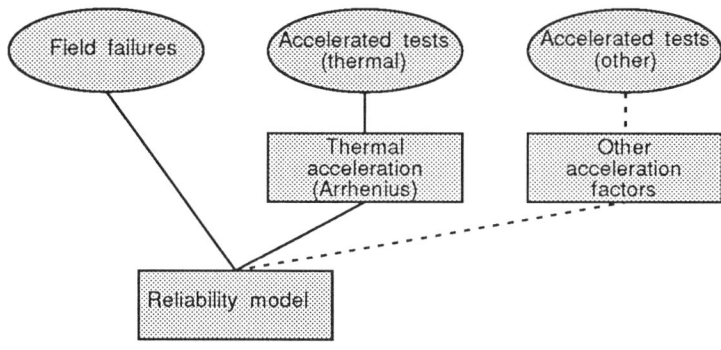

Figure 2.1: Sources of reliability data

Using these two feedback techniques, field failure analysis and accelerated life tests, it is possible to make reliability prediction models for many electronic components. Only in those cases where failure mechanisms are not governed by chemical degradation effects will it be necessary to find other acceleration factors usable in accelerated life tests. (See figure 2.1) At this moment there are a number of organizations working on the development of standard component reliability prediction models. These organizations are mainly to be found in those branches of the electronics industry where reliability is traditionally important. Well-known reliability prediction handbooks (with the related organizations) are given in the table below.

Handbook	Source	Aims
MIL-HDBK-217 [MIL87]	United States Department of Defence (Rome Air Development Centre)	Military (Commercial)
BT-HRD [BT87]	British Telecom	Telecommunication

Beside these standard handbooks many component manufacturers have company- internal reliability data with related component models. The main purpose of all these handbooks is to compare the reliability of different circuits (often in the early design phases of such a circuit) on the basis of component reliability.

The following sections will discuss the common aspects of most reliability prediction handbooks and will discuss some of the best known reliability prediction handbooks with the basic types of reliability models used. As many additional models were developed for the reliability prediction of integrated circuits these IC models will be discussed in a separate section.

2.2.1. *Statistical evaluation methods used in the development of reliability prediction models*

In the previous chapters the reliability of a circuit or system was defined as

The probability R(t) of a circuit or system to perform its intended   (2.2)
function for a given period of time.

It is possible to define this statement (for a large number of identical circuits, used under similar circumstances) in way using the expression:

$$R(t) \;=\; \frac{\text{nr. of circuits operational at time} \;=\; t}{\text{nr. of circuits operational at time} \;=\; 0} \;=\; \frac{n(t)}{n(0)} \qquad (2.3)$$

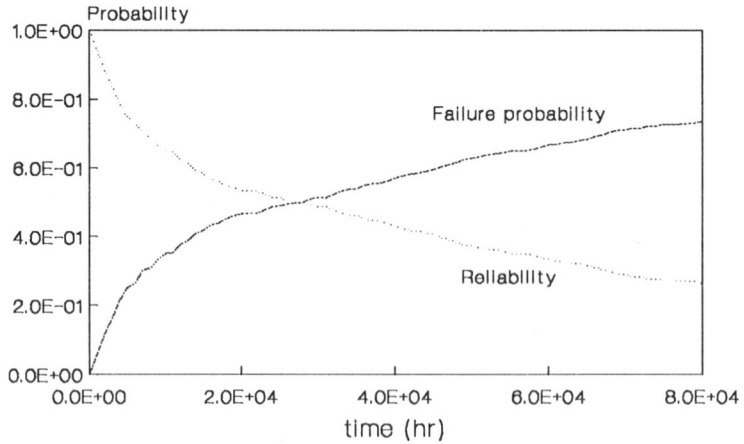

Figure 2.2: Practical failure data and probabilities

A practical example illustrating the use of this definition is given in figure 2.2. From these data it is possible to derive the reliability or the failure probability as a function of time. The failure probability $F(t)$ is defined as:

$$F(t) = 1 - R(t) \qquad (2.4)$$

Another important parameter, often used in reliability analysis, is the failure rate $\lambda(t)$:

$$\lambda(t) = \frac{1}{\Delta t} \frac{\text{nr. of failures }(t..t+\Delta t)}{\text{nr. of systems }(t)} \quad \textit{(failures/hour)} \tag{2.5}$$

The relation between $\lambda(t)$ and R(t) can be expressed using

$$\lambda(t) = \frac{-d\,R(t)/dt}{R(t)} \tag{2.6}$$

Many traditional reliability models are based on the assumptions that it is possible to consider circuits within a large batch as identical and that actual failures occur only at random. Random failures are considered those failures with a constant, time-independent, failure rate.

$$\lambda(t) = \lambda_{constant} \tag{2.7}$$

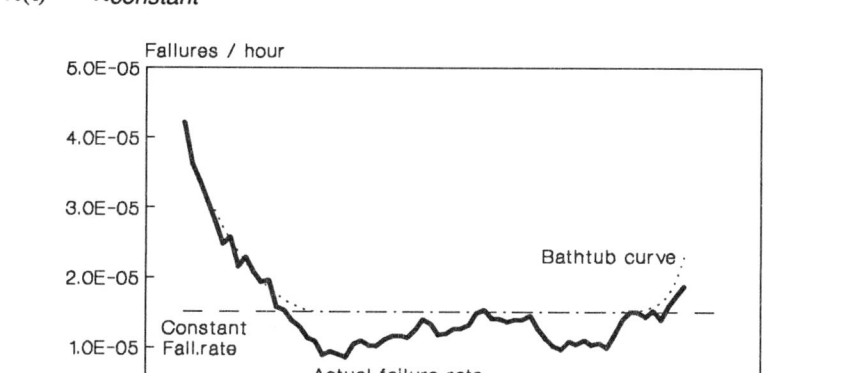

Figure 2.3: Constant failure rate approximation

For a batch of identical circuits with only[*] random failures it is possible to derive that

$$R(t) = e^{-\lambda t} \tag{2.8}$$

---

[*]   Often this constant curve is corrected with an increased failure in the early life of a system and an increasing failure rate at the end of a systems useful life. This results in the so-called "bathtub" reliability curve.

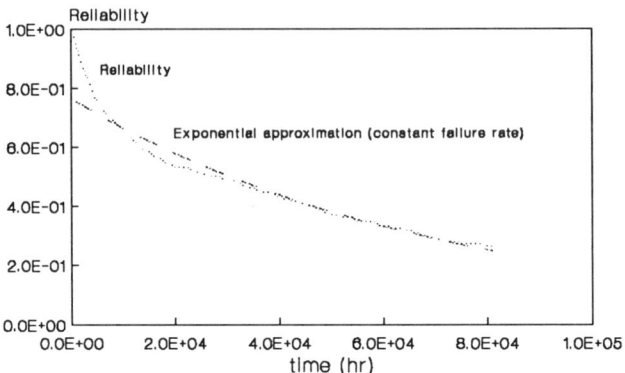

Figure 2.4: Actual and approx. reliability

Quite often electronic circuits are considered as series systems in respect to reliability; all components within a system are considered necessary for a functional correct system. Also failures within a circuit are considered random independent failures[*]. This leads to the following expression:

$$R_{system}(t) = \prod_{i=1}^{n} R_{component\ i}(t) \tag{2.9}$$

Combined with the previous equations this leads to the following, well-known, definition of system reliability based on component reliability:

$$R_{system}(t) = \begin{cases} \displaystyle\prod_{i=1}^{n} R_{component\ i}(t) & = e^{-t\sum_{i=1}^{n}\lambda_i} \\ \lambda_i(t) = \lambda_i\ (romanconstant) \end{cases} \tag{2.10}$$

Using this equation it is possible to calculate the reliability for a circuit or system using part failure rate figures. Although many people have found that this approach is in many respects not realistic, this serial/exponential reliability model is nowadays (often because of its simplicity) used in almost every reliability analysis method [Jen89] [Pec88]. It is important that this model is only valid under the assumptions made above; the exponential reliability model is valid[**] only for:

---

[*]   The only inter- dependencies taken into account are often common stress factors such as environmental temperature.

[**]   In practice there will always be differences between individual circuits and not all failures can be assumed random failures.

— large batches of circuits,

— the circuits have to be perfectly identical,

— failures only occur at random.

In accordance with this list the traditional way to verify or measure the reliability of a certain system is to have a large batch of identical systems operating under identical, well defined environmental conditions and to check the number of surviving systems at a certain time. A disadvantage of this method is that the resulting reliability figure is only relevant for the environmental conditions mentioned above. Often reliability analysis is used not only for the purpose of analyzing the reliability of one single system but also to compare the reliability of different systems under identical environmental conditions or of similar systems under different environmental conditions. Another possibility is to use reliability analysis results for the reliability prediction of new (but similar) systems. For these purposes a wide range of reliability handbooks are used. Reliability handbooks intend to present independent, general applicable reliability models, usable for a wide range of electronic parts. The following sections will present some well-known reliability handbooks, the models used in these handbooks and some of the constraints related to the use of these handbooks. As, mainly for integrated circuits, more detailed reliability models exist, some of these models will be discussed in a separate paragraph on IC reliability.

2.2.2. *United States Department of Defence MIL-HDBK-217*

The MIL-HDBK-217 establishes uniform methods for predicting the reliability of electronic parts and systems. It provides a common basis for reliability predictions during acquisition programs for military electronic systems and equipment. It also establishes a common basis for comparing and evaluating reliability predictions of related or competitive designs.

The handbook contains two methods of reliability prediction: "Part Stress Analysis" (PSA) and "Parts Count Analysis" (PCA). These methods vary in the degree of information needed to apply them. PSA requires the greater amount of detail information and is used later in the design process when circuit structures are known. PCA only requires part quantities, quality levels and the application environment and is therefore less accurate but useful during the early design phase and during the proposal formulation or "tentative device specification".

A similarity between PCA and PSA is that both prediction techniques use the same formulas. For PCA only estimated values are used where PSA uses calculated or measured values. The formulas are obtained by combining practical and theoretical models. There is a similarity between most part models. Generally speaking a failure rate formula will look like:

$$\lambda_p = \lambda_b \, \pi_e \, \pi_q \, \pi_{..}$$ (2.11)

λ_p = part failure rate.

λ_b = base failure rate, dependent on temperature and applied stress.

$\pi_{..}$ = acceleration factors for the used environmental application and other parameters that will affect the part reliability. For example:

π_e = Environment

π_q = Quality

Environment and quality are used for most parts, other π factors are part dependent. The environmental acceleration factors are defined for a number of standard conditions. These standard conditions are varying from laboratory conditions to a cannon bullet during launch. Because of the fact that the MIL-HDBK 217 was developed for military purposes most standard environments and most standard quality levels are according to military standards, but the MIL-HDBK 217 contains also information to predict failure rates for non-military environments and quality levels.

The base failure rate is usually expressed by a model relating the influence of electrical and temperature stresses on a part. One of the most characteristic parts of many models is the relation between temperature and failure rate. These models use thermal stresses in a form related to the Arrhenius Law. Another stress parameter for many devices is related to the dissipated power. Often this parameter is used to describe the thermal effects of power dissipation on the lifetime of a circuit. In many cases the thermal and power effects are combined in one term of the failure rate formula. The following formula gives a typical example of mixed power-thermal influence.

$$\lambda_b = A.e^{\left(N\frac{t}{T_{amb}+T_{power}} + \left(\frac{T_{amb}+T_{power}}{T}m \right)^P \right)}$$ (2.12)

This formula, describing the reliability of discrete semiconductors, consists of Arrhenius law with a correction factor for a decreasing activation energy at higher temperatures. For discrete transistors the corresponding activation energy is, for example, about 0.9 eV.

Other acceleration factors are modelled in terms of acceleration factors $\pi_{...}$. The data used to model these acceleration factors are obtained by means of analysis of field failures and manufacturers data. Using this method it is possible to model the effects of using a component under certain environmental conditions, the effect of using certain methods of component quality screening, etc. The table below gives a survey of the most common MIL-HDBK-217 parameters with their meaning, their acceleration factors and their origin.

Param.	Description	Influence factors	Source
	All parts		
π_e	Environmental acceleration factor	environment class	heuristic
π_q	Quality acceleration factor	component screening	heuristic
	Integrated circuits	$(\lambda = \pi_q \, \pi_l \, (C_1 \, \pi_t \, \pi_v \, \pi_{pt} + (C_1 + C_3) \, \pi_e)$	
π_t	Thermal acceleration factor	thermal, device struct	Arrhenius, heuristic
π_v	Voltage derating factor MOS only	application class	heuristic
π_{pt}	Correction factor for programming technique ROM/PROM only	device structure	heuristic
C_1, C_2	Complexity factor, depending on number of bits/gates	device structure	heuristic, (thermal conductivity package)
C_3	Complexity factor, depending on package	device structure	heuristic (thermal conductivity package)
π_l	Learning factor	device structure, depending on the maturity of the production process of a part	heuristic
	Discrete semiconductors	$\lambda = \lambda_b \, \pi_e \, \pi_a \, \pi_q \, \pi_r \, \pi_{s2} \, \pi_c$	
λ_b	Basic failure rate	thermal, device structure, (application class)	Arrhenius, heuristic
π_a	Application factor	application class	heuristic
π_r	Rating correction factor	application class	heuristic
π_c	Complexity factor	device structure	heuristic
π_{s2}	Voltage stress correction factor	application (mean Vce)	heuristic
π_f	Frequency/peak power correction factor microwave transistors only	application class	heuristic
	Resistors	$\lambda = \lambda_b \, \pi_{taps} \, \pi_r \, \pi_v \, \pi_c \, \pi_e \, \pi_q$	
λ_b	Basic failure rate	thermal, device structure, (application class)	Arrhenius, heuristic
π_r	Resistance value correction factor	device structure	heuristic, (area and thickness of material)
π_c	Construction correction factor	device structure	heuristic

Param.	Description	Influence factors	Source
π_v	Voltage correction factor variable resistors only	application class	heuristic
π_{taps}	Tap connection correction factor variable resistors only	device structure	heuristic
	Capacitors	$\lambda = \lambda_b \, \pi_{cv} \, \pi_{sr} \, \pi_q \, \pi_c$	
λ_b	Basic failure rate	thermal, device structure, (application class)	Arrhenius, heuristic
π_{sr}	Series resistance value correction factor elcos only	device structure	heuristic
π_{cv}	Capacitance value correction factor	device structure	heuristic, (area of cap. plates)
π_c	Construction correction factor	device structure	heuristic

Most other parts discussed in the MIL-HDBK-217 are similar to the models used to describe the reliability of resistors and capacitors.

Most parameters used in the MIL-HDBK-217 models are either related to device structure or the application of the component. Especially the application bound parameters have the form of correction factors, derived from heuristic data; no physical failure mechanisms are involved.

2.2.3. *British Telecom HRD-4*

The British Telecom Handbook of Reliability Data is quite similar in approach to the MIL-HDBK-217. In case of the British Telecom handbook the term Part Stress Count analysis replaces the term Part Stress Analysis used in the MIL-HDBK-217. The British Telecom Handbook of Reliability Data uses formulas of the form:

$$\lambda_p = \lambda_b \, \pi_t \, \pi_q \, \pi_e \qquad\qquad (2.13)$$

λ_p = part failure rate.

λ_b = base failure rate.

π_t = Thermal acceleration factor.
π_q = Quality correction factor.
π_e = Environmental acceleration factor.

For some components not all these parameters are used. For those components where π_t is used, it takes the form of the Arrhenius law. For integrated circuits two degradation processes are used; one with an activation energy of 0.3 eV and another with an activation energy of 1 eV. For most components the λ_b is used as a constant factor, independent of external stress factors. The table below gives an overview of the most common BT-HRD-4 parameters with their meaning, their influence factors and their origin.

Param.	Description	Influence factors	Source
	All parts		
π_e	Environmental accelera-tion factor	environment class	heuristic
π_q	Quality correction factor	component screening	heuristic
	Integrated circuits		
λ_b	Base failure rate, de-pending on number of bits/gates	device structure	heuristic
π_t	Thermal acceleration fac-tor $$A.e^{\frac{-3500}{T_j}} + e^{\frac{-11600}{T_j}}$$	thermal, device structure	Arrhenius (2 fail. mech.), heuristic
	Discrete semiconductors and passive components		
λ_b	Base failure rate	device structure	heuristic

Generally speaking the BT-HRD-4 is less detailed than the MIL-HDBK-217. It is possible to explain many of the differences between both handbooks by means of regarding the backgrounds of the used data. The MIL handbook aims at the military world and needs, for that purpose, to cover a wider range of environmen-tal an application influence factors compared to the British Telecom handbook which only covers equipment for telecommunication purposes.

Most other handbooks use models for most components equal or similar to either the MIL- or the BT- handbook. Only for integrated circuits do some other, more detailed, models exist [Wal89].[*] The following section will discuss these failure rate models, especially where differences with existing models exist.

---

[*] Research on existing reliability models, especially for integrated circuits, was carried out by Ir. J.S. Walinga as MSc assignment as part of this project.

2.2.4. *Other reliability models for integrated circuits*

2.2.4.1   *The Philips failure rate model for integrated circuits*

The Philips model is a failure rate model for internal use, in which two types of integrated circuits are modelled [Phi88]. The amount of integration is modelled as follows:

— SSI/MSI devices (small scale and medium scale integrated circuits)

— LSI/VLSI devices (large scale and very large scale integrated circuits)

The first part of the well known bathtub failure rate approximation is described partly in the Philips model. For integrated circuits it has (according to the Philips data sheet) the following structure.

Param.	Description	Influence factors	Source
	$\lambda_p = \pi_S\ \pi_T\ \lambda_e$ $\lambda_p = \pi_S\ \pi_T\ \lambda_c$	t<300h t>300h	
λ_e λ_c	Base failure rate (t<300h) Base failure rate (t>300h)	device structure	heuristic
π_S	Size acceleration factor	device structure (complexity)	heuristic
π_T	Thermal acceleration factor	thermal, device structure	Arrhenius, heuristic

In the Philips model the failure rate is independent of the stress behaviour. Thus the only application related parameter influencing the failure behaviour is the temperature.

2.2.4.2   *Kasouf and Mercurio model for integrated circuits*

The Kasouf and Mercurio model is similar to, but less complex than, the MIL-HDBK-217 model. Kasouf and Mercurio propose a model for random logic devices based upon gate and bond complexity [Rey82].

Param.	Description	Influence factors	Source
	$\lambda_p = \pi_q\, C\, e^{\left[A\left(\frac{1}{T_R} - \frac{1}{T_J} \right) \right]}$ $C = K + 6.0\ x\ 10^{-5}\ N_p + 9.25\ x\ 10^{-6} G$		

Param.	Description	Influence factors	Source
K	Constant	device structure	heuristic
G	Number of gates	device structure	
N_p	Number of pins	device structure	
C	Complexity factor	device structure	heuristic
T_R	Rated temperature	device structure	
π_q	Quality acceleration factor	component screening	heuristic
A	constant	E_a/k	

2.2.4.3 *The CNET failure rate model for integrated circuits*

An other model for failure rate prediction is developed and published by the French Centre National d'Etudes des Telecommunications (CNET model). This model is very similar to that of the MIL-HDBK-217 E, which was mentioned in one of the previous paragraphs. The equation of the CNET model is:

Param.	Description	Influence factors	Source
	$\lambda_p = \pi_q (C_1\, \pi_T\, \pi_V + (C_2 + \pi_P)\, \pi_e)\, \pi_L$		
π_q	Quality acceleration factor	component screening	heuristic
π_T	Technology factor	device structure	Arrhenius (2 degr. mech)
	$$e^{\left[11606\, \varphi \left\{ \frac{1}{T_{r1}} - \frac{1}{T} \right\} \right]} + e^{\left[11606\, \varphi \left\{ \frac{1}{T_{r2}} - \frac{1}{T} \right\} \right]}$$		
π_e	Environmental acceleration factor	environment class	heuristic
π_P	Package factor	device structure	heuristic (thermal conductivity package)
C1, C2	Complexity factors, depending on the number of bits/gates	device structure	heuristic

The difference between the MIL-HDBK-217E model and that of the CNET model is the fact that in the CNET model two different thermal acceleration mechanisms are described and in the MIL-HDBK-217E only one. These two failure mechanisms have two different activation energy levels; for example 0.3 eV and 1.0 eV for bipolar ICs. They are taken into account in the technology factor.

2.2.4.4 *The extended MIL-HDBK-217E model for integrated circuits (Pantic model)*

This model is developed by Dragan M. Pantic to take more effects in the failure rate model [Pan84]. In the general model of the MIL-HDBK-217E the failure rates are the same for the total period of manufacturing. Practical by Pantic have learned that the failure rates decreased in course of time and also that the manufacturers affect the failure rates, so two new acceleration factors are added to the general MIL-HDBK-217 E formula. In general the formula for integrated circuits is:

Param.	Description	Influence factors	Source
$\lambda_p = \pi_Q\,\pi_D\,\pi_M\,(\,C_1\,\pi_T\,\pi_V + C_2\,\pi_E\,)\,\pi_L$			
π_D	Data code factor representing the maturity of technology in general	device structure (technology)	heuristic
π_M	Manufacturers factor representing the maturity of the specific manufacturer	device structure (production)	heuristic

All the other factors are equal to the normal MIL-HDBK-217 E standard. With this model the decrease of the failure rate in course of the time is described. According to Pantic this means that the quality increases the longer a circuit is manufactured.

The failure behaviour is the same as that of the other failure rate model. The six failure prediction methods described by these six models have in general the same structure. The part failure rate depends mainly on the temperature.

2.2.5. *Summary of standard failure rate handbooks*

Although the individual formulas describing the failure rate are quite different among the various reliability analysis handbooks they have several aspects in common. First of all the relation between part failure rate and effective device temperature[*] is expressed in various forms of the Arrhenius law. Activation energies used in this expression are often very different for similar components. (See one of the following paragraphs for a practical example.) Most other influence factors are modelled in the form of acceleration factors. The acceleration factors are nearly always presented in the form of tables divided in certain

---

[*]   The term effective temperature rating is used to describe the summarized effects of both internal power dissipation and environmental temperature on the device.

classes. Most times these classification tables are based on practical experience and do not use an underlying physical model.

It is possible to split the influence factors in five main categories:

— Parameters related to the application class

— Parameters related to the effective device temperature

— Parameters related to the standard environment class

— Parameters related to the device structure

— Parameters related to the used component screening techniques

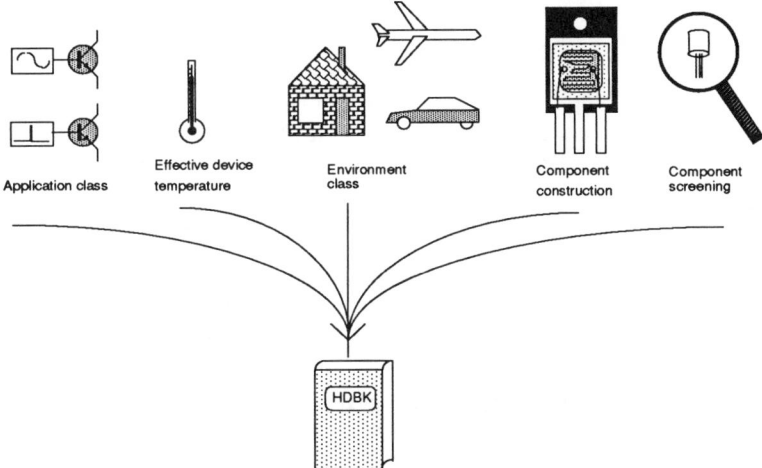

| Application class | Effective device temperature | Environment class | Component construction | Component screening |

Figure 2.5: Sources of reliability data

Especially for integrated circuits many models use extensions to the common used MIL handbook model, taking more detailed effects related to manufacturing and device complexity into account.

The majority of the acceleration factors is either related to the effective device temperature, the device structure or is environment bound. The application is only introduced using an application correction factor or, via the thermal effects of power dissipation, as a term in the Arrhenius law. In none of the discussed failure rate handbooks data considering the application are used in more detail.

The following section will present reliability predictions for some practical circuits. The resulting failure rate figures are compared to the results of practical experiments. As a next step the results of the reliability prediction are used to derive possible suggestions for reliability improvements. The following sections will concentrate especially on several semiconductors as the differences between the reliability prediction results (and the related optimization suggestions) show a vast difference with the results of practical experiments.

2.3. Reliability calculations using standard reliability prediction handbooks

To test the usability of standard reliability prediction handbooks in the process of reliability optimization the following practical circuits * were used:

— Circuit A: self oscillating power supply.

— Circuit B: self oscillating power supply.

— Circuit C: motor drive circuit.

These three circuits will be used to illustrate the theory presented in this book. All circuits were used in Philips video cassette recorders. The practical failure rate figures, used for verification purposes, are based upon field data obtained during the period 1987-1989. The examples will focus on certain components as these components will be used for a more detailed analysis in the following chapters. The circuits were selected because company experience showed that reliability problems were expected especially in power circuits. The examples will show only the most important parts of the circuits. The actual calculations and the verification measurements were performed on the entire circuit.

2.3.1. *Reliability prediction of test circuits compared to practical reliability figures*

Before using one of the reliability prediction methods described earlier in this chapter in a reliability optimization process it will be necessary to compare the reliability prediction figures of the different prediction methods with each other and to compare the predicted failure rate figures with figures obtained from practical situations. First the reliability figures of both power supplies are discussed. For these circuits failure rate predictions were made using the MIL-HDBK-217 and the British Telecom handbook HRD-4. As the third circuit contains an IC this circuit, and especially reliability aspects of the IC, will be discussed in more detail.

---

*   A short explanation of the used circuits and components is given in appendix A.

The following paragraphs will concentrate mainly on the results of the reliability prediction; Appendix A of this book will describe the circuits and their reliability in more detail.

2.3.1.1 *Reliability prediction test circuits A and B*

Figure 2.6 gives in detail the failure rate figures of circuit A and B in accordance with MIL-HDBK-217E and BT-HRD-4 combined with actual reliability figures. It is obvious that wide differences in failure rate numbers exist for most components. Generally speaking the MIL-HDBK-217 prediction data tends to be worse (higher value) than the BT-HRD-4 figures. The BT-HRD-4 figures in turn tend to be worse than the practical figures. To check whether the failure rate ratio for the components is similar according to the different failure rate methods the relative failure rate figures are given in figure 2.7. This figure shows especially the relative failure rate for the components with the highest practical failure rate.

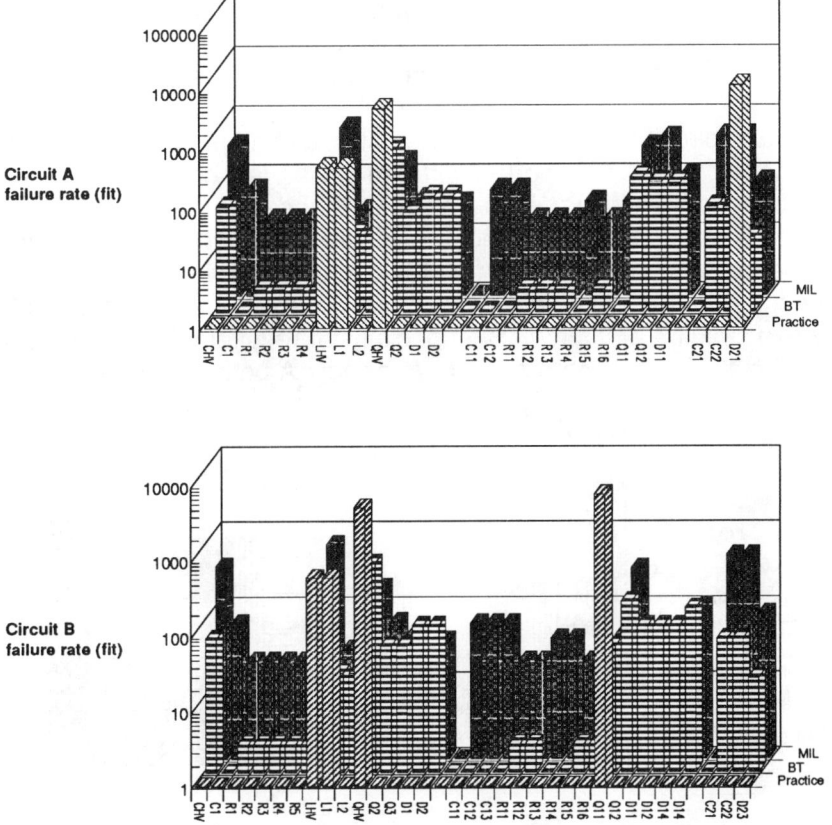

Figure 2.6: Failure rate circuits A and B

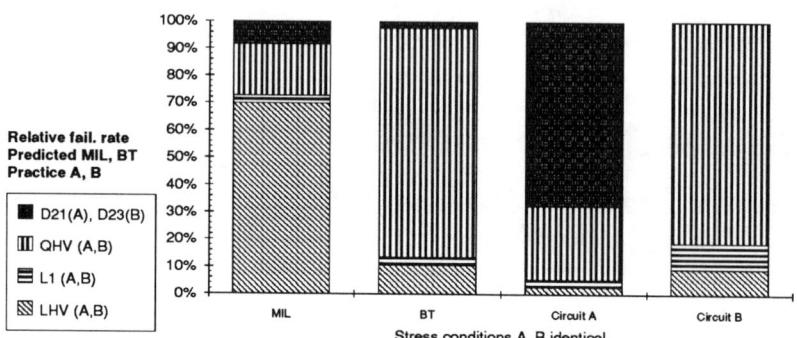

Figure 2.7: Failure rate circuits A and B

An important conclusion from this figure is that not only the absolute failure rate figures but also the relative failure rate figures show vast differences. Especially interesting are the differences for D21 in circuit A and transistor QHV in both circuits between the predicted figures and practical figures.

As circuit A and circuit B are very similar in structure it also becomes interesting to check whether there are important differences in practical failure rate for those components used both in circuit A and circuit B under (for MIL and BT) identical conditions. It is obvious that the predicted failure rate under these conditions will be the same for these components for both circuit A and circuit B. See figure 2.8. One of the most interesting points in this figure is the difference in practical failure rate between circuit A and circuit B, especially for rectifier diodes D21 (circuit A) and D23 (circuit B). As can be seen in Appendix A both components are used

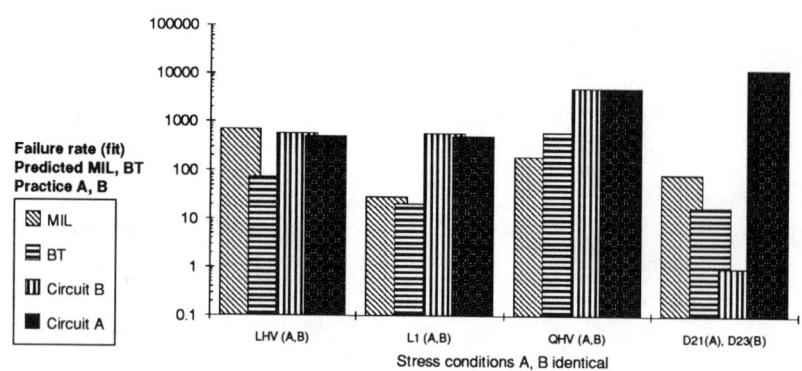

Figure 2.8: Comparing identical components in A & B

under (for MIL and BT) exactly the same conditions. Diode D21 in circuit A and diode D23 in circuit B are physically the same diode. The difference in practical failure rate is greater than a factor 10^4.

The first conclusion from this paragraph is the necessity for a detailed explanation of such differences before it is possible to use one of the discussed failure rate prediction methods in a design optimization process.

2.3.1.2   *Reliability prediction test circuit C*

For the single components of circuit C the same method is used as described above. For integrated circuits many more reliability models exist. Therefore for the integrated circuit five reliability predictions were made. Reliability figures for circuit C are presented in figure 2.9.

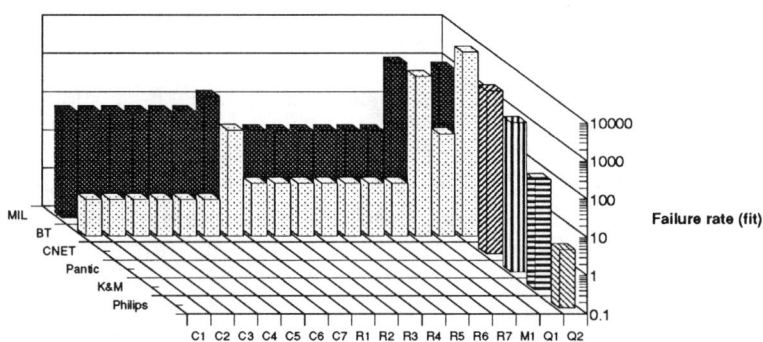

Figure 2.9: Failure rate circuit C

As with the previous circuits circuit C also shows important differences in failure rate figures for the different prediction methods. Especially for the integrated circuit Q2 the various prediction methods disagree over a wide interval. Difference between the Philips model and the MIL model is about 10^3. To check different influence factors in the different prediction models a more detailed prediction was made for this IC. As, in most failure rate models for integrated circuits the temperature is one of the most important parameters a separate failure rate prediction was made for standard temperature (30 °C) and for a higher temperature (50 °C)[*]. This to check if there are any differences in temperature

---

[*]   Many failure rate handbooks define a reference ambient temperature of 30 °C for standard calculations.

influence between the different models. As some of the failure rate models have separate figures for so-called *early failures* these early failures are also considered separately. Results of these calculations are given in figure 2.10.

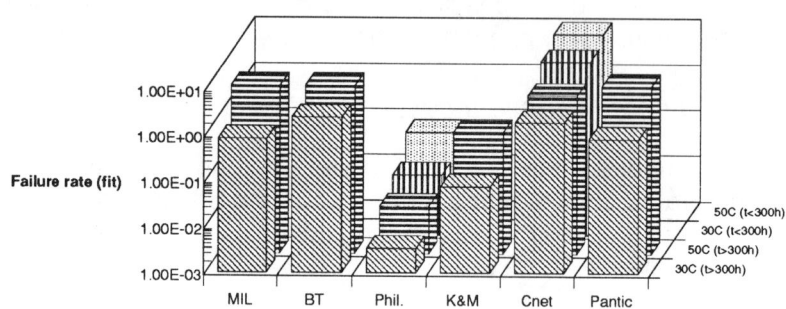

Figure 2.10: Influences of temperature and time < 300h

This figure shows that for the same integrated circuit many different failure rate predictions are possible. The differences in prediction result are due to fundamental differences in the models themselves. The only similarity between the models is the fact that they all use some form of the Arrhenius law. The most important term in this law, the activation energy of the degradation process, is given for these models in figure 2.11.

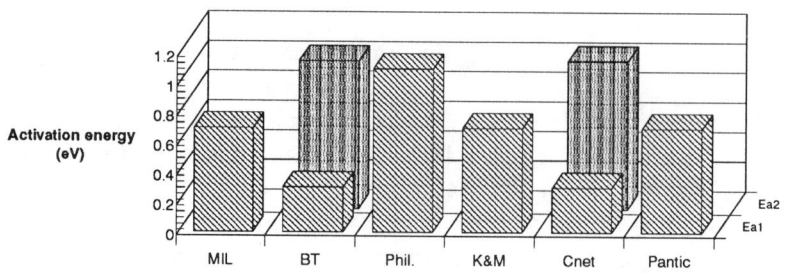

Figure 2.11: Different activation energies for ICs

It is obvious that there are important differences in activation energy for the different failure rate prediction models. Another difference is the fact that some models use one single degradation process while other models use two degradation processes with two activation energies. It is obvious that not only in resulting value but also in model structure there are differences between the different failure rate models.

2.3.2. *Summary of practical reliability predictions*

In the previous sections reliability predictions were made for three practical circuits. For these predictions different failure rate prediction methods were used. For single components predictions were made in accordance with MIL-HDBK-217 and BT-HRD-4, as these handbooks use the two most common used prediction models for single components. For integrated circuits many more reliability prediction models exist. Therefore in case of the failure rate prediction of an integrated circuit six prediction models were used. Where possible prediction results were compared to field failure rate figures.

From the previous sections it is possible to derive the three following conclusions:

> *It is, for most electronic components, not possible to derive one* (2.14)
> *comprehensive component failure rate prediction model from*
> *the discussed existing failure rate models; the discussed models*
> *are often not only different in result but also in model structure.*

> *The only common factor in the discussed failure rate models is* (2.15)
> *the Arrhenius law; the differences in used activation energies,*
> *however, indicate the use of fundamentally different degradation*
> *mechanisms for the same components.*

> *The significant difference in practical failure rate for a component* (2.16)
> *used in two similar but different circuits under, for the existing*
> *failure rate prediction models, identical circumstances shows*
> *considerable deficiencies in interaction mechanisms described*
> *in the existing failure rate prediction models.*

Although these statements make the use of existing reliability prediction methods in a design-reliability optimization process very unlikely, the following section will discuss theoretical possibilities for such an optimization.

2.4. Reliability optimization according to standard reliability prediction handbooks

Although the previous sections showed some important problems in existing standard reliability prediction models this section will try to use the interaction mechanisms described in the presented handbooks for the purpose of reliability optimization. First of all it will be necessary to define where design optimization is possible.

> *The process of optimizing a circuit will consist of (often step-*   (2.17)
> *wise) changing the design structure or design parameters until*
> *an optimal structure in respect to a set of target parameters has*
> *been achieved.*

These target parameters are: functional parameters, often presented of a set of target specifications; cost parameters, often presented in the form of a target price and a set of reliability and quality aims. Quite often constraints in one of these parameter sets will limit the degree of freedom allowed in other parameter sets. It is, for example, hardly practical to use an environment class parameter in a design optimization process. If a system is intended for use in a car it will hardly ever be possible to modify the environmental conditions in such a way that, for example, conditions coming close to laboratory conditions are reached. For reliability optimization as part of the greater process of design optimization it will be necessary to express reliability parameters in terms of designable parameters, or, in more detail:

> *Reliability optimization as integrated part of the design process*   (2.18)
> *of electronic circuits and systems requires reliability demands*
> *expressed in terms of designable parameters.*

> *Designable parameters are those parameters where the de-*   (2.19)
> *signer has certain degrees of freedom to introduce changes.*

Only in such a way will a comprehensive optimization process be possible. The following sections will discuss the possibility of using standard reliability prediction models for reliability optimization. First the prediction models are checked for the existence of usable (designable) parameters while in the following sections the theoretical effect of such a reliability optimization will be discussed.

2.4.1. *Parameters in existing reliability models usable for reliability optimization*

As discussed earlier in this chapter, reliability optimization requires models giving a direct relation between changeable design parameters and the reliability of a system. All the existing reliability prediction models use parameters which can be divided in five different categories. Figure 2.12 gives the relation between these categories and the actual sources of reliability problems.

A first conclusion from this figure might be that all problem sources are related to at least one of the mentioned parameter categories. The problem remains that in many cases the models use heuristic classification parameters instead of detailed relations. The question is: are these classifications usable in an optimization process? The table below gives some suggestions which may lead to reliability improvements, both according to MIL and BT.

Part number	Optimization possibility	Constraints
All parts	Part quality-class screening	Possible
	Decrease temperature	May result in conflict with spec. demands.
	Change application environment	Often impossible due to spec. demands

Problem with these optimization criteria is that for many applications they are hardly applicable to practical situations. The only possible practical solution derived from the table above is the possibility to use screening techniques to improve the quality of the used parts and the possibility to decrease in some cases the part temperature. It is also possible to obtain design improvements by means of changing application environment or other environment related parameters but, generally speaking, these possibilities result in conflicts with the demands, given in the design specifications of a circuit. It is, for example, hardly possible to use consumer electronics under laboratory conditions only.

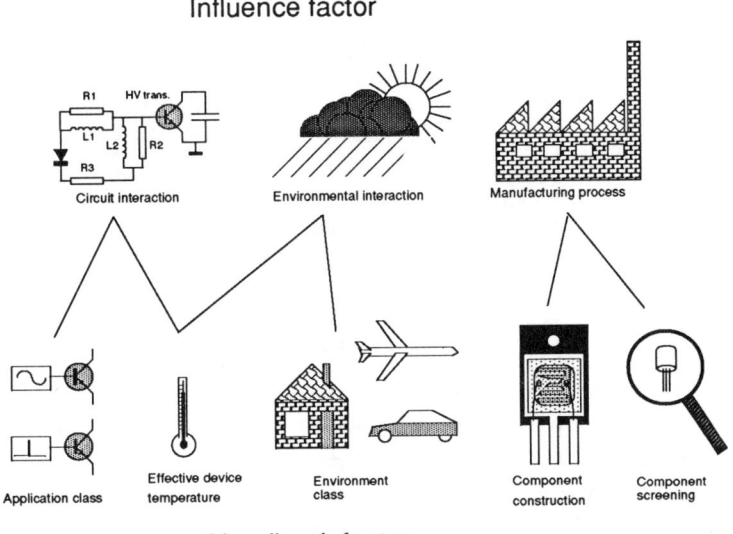

Figure 2.12: Sources of reliability data

The table below gives some more detailed optimization possibilities, all according to the MIL-HDBK-217. The improvements given in this table are all application related.

Part number	Circuit	Optimization possibility	Constraints
Integrated circuits	(A), C	Decrease junction temperature (Power load)	Impossible due to spec.
Discrete transistor	A, B, C	Decrease power load	Impossible due to spec. demands
		Change to equivalent component with higher rating	Possible
		Decrease voltage load	Impossible due to spec. demands
		Change to equivalent component with higher rating	Possible
Secondary rectifier diode	A,B	Decrease power load	Impossible due to spec. demands
		Change to equivalent component with higher rating	Possible
Secondary rectifier diode (cont)		Decrease reverse voltage load	Impossible due to spec. demands
		Change to equivalent component with higher rating	Possible
Capacitors	A, B, C	Decrease voltage lo ad	Impossible due to spec. demands
		Change to equivalent component with higher rating	Possible
Resistors	A, B, C	Decrease voltage load	Impossible due to spec. demands
		Change to equivalent component with higher rating	Possible
Inductors		No direct specification-related improvements possible.	

One of the most important conclusions of the table above is the fact that the only practical solution appears to be the selection of a "bigger" part. The term bigger relates for all components to the effective temperature rating. For some components correction factors are introduced for the application category. This leads to the following conclusion:

> *Standard failure rate prediction handbooks use models in which* (2.20)
> *effective device temperature is the dominant circuit related*
> *stress influence factor. If other circuit related stress parameters*
> *are used they are used in the form of correction factors.*

2.4.2. *Reliability optimizations derived from standard reliability prediction models*

As demonstrated in the previous section the existing reliability prediction models offer only limited possibilities for reliability optimization. This paragraph will discuss the effect of introducing such optimizations. This will be illustrated using some practical examples.

One of the most detailed models, in respect to describing interaction mechanisms, is the MIL-HDBK-217 model for single semiconductors. This model uses dissipated power and environmental temperature to calculate the effective device temperature and introduces a correction factor for applied Vce. The figure below shows the effect of reducing these stress parameters with a factor 0.5. Figure 2.13 shows the effects of such a stress reduction on QHV in circuit A and B and diodes D21 (circuit A) and D23 (circuit B). Although such an extreme stress reduction will hardly ever be possible in a practical situation the effect remains small compared to the variations between the MIL model and other (theoretical and practical) reliability figures.

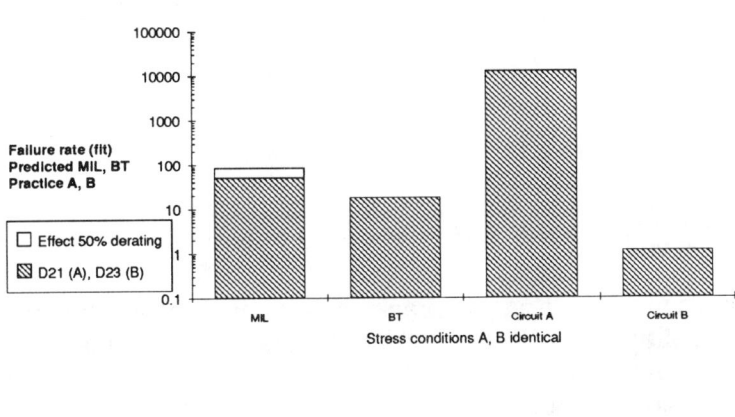

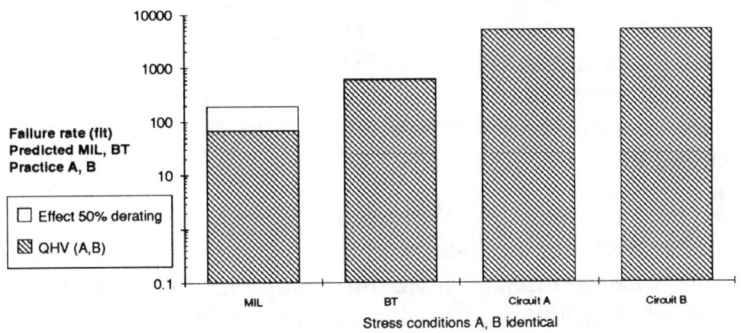

Figure 2.13: MIL optimization effects; 50% derating

The same experiment was repeated for integrated circuit Q2 of circuit C. In this case the junction temperature is the only possible parameter were reliability optimization seems possible. Figure 2.14 shows the theoretical effect of decreasing the junction temperature with 20 °C. In most practical applications such a temperature reduction will hardly be possible. Conclusions from this figure are first that the effects of junction temperature are very different for the used prediction models and second that the predicted reliability improvement remains small compared to the variations among the different models.

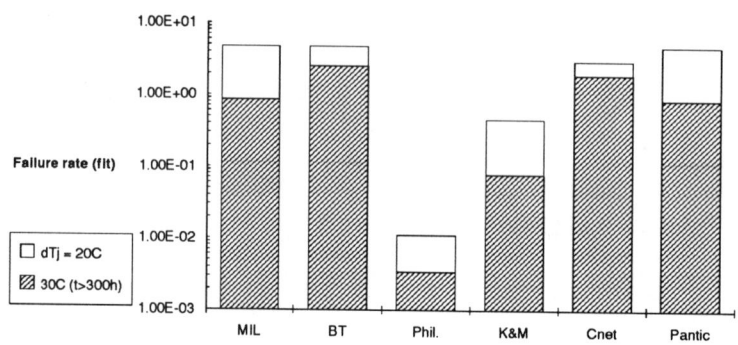

Figure 2.14: MIL optimization effects; thermal

2.4.3. *Summary of reliability optimization using existing reliability prediction models*

The important conclusion from the previous sections is the fact that relative large differences exist among the different failure rate prediction methods. Also the practical figures show quite remarkable differences with the prediction results. Another conclusion from the previous sections is the fact that in many cases it is hardly possible to derive reliability optimization criteria from standard reliability prediction handbooks. Many suggested reliability improvements are hardly practical or will result in conflicts with specification demands. In those cases where reliability improvements seem possible the resulting reliability improvement remains within the uncertainty of the failure rate figures. This results in the following statements:

> *Most existing reliability prediction models give a relation be-*   (2.21)
> *tween a wide range of influence classes and the expected*
> *reliability of a component. Many of the parameters used in these*
> *models are not usable as parameters in a design optimization*
> *process.*

In those cases where reliability prediction models have parame-   (2.22)
ters usable for optimization the effects of such an optimization
will often remain within the uncertainty limits of the predicted
reliability figure.

This leads to the following conclusion:

Standard reliability handbooks are not usable for reliability op-   (2.23)
timization.

This is a logical consequence of the purpose of reliability handbooks. Main
purpose of a vast majority of the standard reliability prediction handbooks is to
provide a common base to compare the influence of wide-range influence factors
such as different application environments with the failure rate of components.
The data used to obtain these figures come from such a wide range of applications that more detailed influence mechanisms are obscured. Although these
wide-range models are very useful for many purposes they miss details required
for optimization purposes. A possible solution seems to be the extension of the
existing reliability models with the description of more detailed effects. Unfortunately it is hardly possible to describe the wanted detailed effects using the
existing feedback techniques. The failure cause analysis is performed at the
moment the component has failed (post-mortem analysis) and at such a moment
it is hard to derive the actual conditions of the circuit at the moment of the actual
failure. Maintaining the wide-range aims of the existing reliability prediction
handbooks and also taking more detailed information of the conditions at the
moment of part failure into account will result in very time-consuming and tedious
feedback processes, but in many cases such wide-range comprehensive models
are not required.

For circuits used under a limited number of environmental conditions it is useful
to restrict the model development to this set of standard conditions. For components used mainly for certain applications it is useful to restrict model development to this function only. This will result in more detailed models, valid only for
a very restricted application working under restricted environmental conditions
but expressing a direct relation between the mentioned problem causes and the
reliability of a system.

All the discussed reliability models predict the failure rate of a given component
in a given circuit as a function of "the stress" on such a component and a number
of external influence factors. For a large batch of "identical" circuits this stress is
based on the average stress on the average component. This is very useful to
compare different classes of components used in different classes of circuits but
provides quite some difficulties when used for optimizing one single circuit.

For many circuits there is not such a thing as an average stress. Quite often the
variations in stress within a single circuit are significant. Also the differences

between individual circuits within a large batch can be considerable. A third point causing variations in reliability is the susceptibility of individual components for certain stress factors. This individual suscpetibility might be quite different for individual components within a large batch. Summarizing: the existing reliability analysis methods do not take differences on the level of individual circuits into account. Some possible differences on this level are

— Differences in stress within a circuit (due to various conditions of use of a circuit)

— Differences in stress between individual circuits within a batch (due to, for example, component tolerances)

— Differences in susceptibility of individual components for the different stress factors

These differences may account for the differences in failure rate for one component used in, for failure handbooks identical circumstances. This especially when the mentioned differences are comperatively large. More detailed reliability analysis should take these effects into account. See figure 2.15.

Influence factor

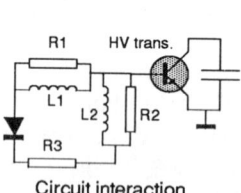  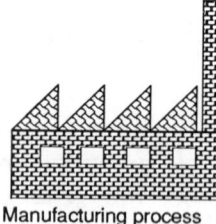

Circuit interaction   Environmental interaction   Manufacturing process

Single circuit parameters

Stress behavior on signal level
Variations in electrical conditions
Variations in environmental conditions

Susceptibility of component failure mechanisms
for the parameters mentioned above

Multiple circuit parameters

Distributions of parameters
important at single circuit level

Effects of parameter tolerances
on stress related signals and
susceptibility of failure mechanisms

Figure 2.15: Parameters required for reliab. optimization

Chapter 3 will describe a method for reliability analysis on the level of individual circuits. The final sections of that chapter will link this new method to traditional methods for reliability analysis.

3

Failure Prediction using Stressor/susceptibility Interaction

3.1. Introduction

Chapter 2 of this book shows different models for the reliability prediction of electronic components. The main purpose of the model discussion is to provide a platform on which reliability predictions can be made during the design phase of circuits and systems. The models should be, according to MIL-HDBK-217, *applicable during the design phases when actual hardware and circuits are being designed* [MIL87]. Main problem with the approach in existing reliability handbooks is that the approach is *top-down*: failure rate models are intended for component *classes* applied in *classes* of circuits under different *classes* of environmental conditions. Although there are considerable differences between the discussed models some aspects are common for all models.

— Existing reliability prediction models predict the failure rate of a given component based on component technology and stress.

— The stress, used in these reliability models, is the mean stress on a component and is nearly always related to thermal effects.

— Individual differences (in stress factors or in susceptibility to stress factors) between circuits and/or components within a large batch are not taken into account.

One of the most interesting results of Chapter 2 was the considerable difference between failure rates predicted in accordance with different prediction methods. Another interesting result was the remarkable difference between predicted and actual failure rates. Apparently the top-down method used in the development of the discussed reliability prediction methods is not suitable for a more detailed analysis. In other words: handbooks like the MIL-HDBK-217 are not able to fulfil their own initial aims in case of a detailed analysis. Therefore another approach will be necessary.

This chapter will discuss a new method for reliability prediction, based on *bottom-up* techniques. While traditional reliability analysis method describe reliability as a function of a number of high-level influence factors this new technique will try to analyze reliability on the level of failure causes for individual components.

The first part of this chapter will present a theoretical method for the prediction of component reliability based on the failure mechanisms in a component. The second part will link this new method to traditional reliability analysis methods. The third and final part shows the relation between reliability of the components on one hand and designable circuit parameters related on the other.

3.2. Part failure mechanisms

Although theoretical models often describe components as ideal unchanging devices practical electronic components are in fact dynamic systems in constant interaction with their direct environment. Due to this interaction it is possible that, after a certain time, a component will no longer be able to fulfil its intended purpose. In this case the component fails. To distinguish clearly between functional correct and failing components it is necessary to define the terms *operational* and *failure*:

> *A system (or component) is considered operational if it is able*   (3.1)
> *to meet fully with its specifications.*

> *A system (or component) fails if it is not able to fulfil one or more*   (3.2)
> *of its specifications.*

The processes which may lead to component failure are called *failure mechanisms*:

> *A failure mechanism is a process in an electronic component*   (3.3)
> *which may lead to failure of this component.*

Part failure mechanisms are related to the way the components are used, the physical structure of the components and other internal and external influences relating to the probability of failure of a component.

To analyze the influences of the various mechanisms on the lifetime of a circuit the term stressor is introduced:

> *A stressor is a physical entity influencing the lifetime of a*   (3.4)
> *component or circuit; A stressor, indicating a physical entity x*
> *will be denoted as* ψ_x.

Combining these two definitions leads to the following relation:

> *A failure mechanism is a physical degradation process of an* (3.5)
> *electronic component influenced by one or more stressors.*

For the analysis of the probability that a component will fail due to such a failure mechanism it is not only important to know the stressors but also to know the susceptibility for these stressors. See figure 3.1. The following sections will derive theoretical stressor- and susceptibility- functions in order to derive a platform for the calculation of component failure probability.

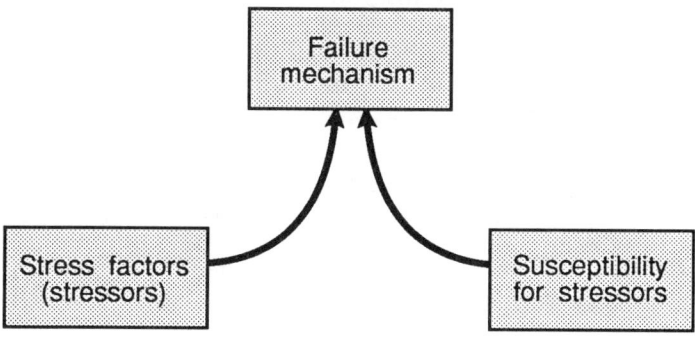

Figure 3.1

3.3. Stressors

For components used in electronic circuits it is possible to split possible stressors into certain groups. First there are electrical stressors; parameters, directly related to the electrical behaviour of a circuit. Second group of stressors are the mechanical stressors; stressors related to the mechanical environment of a component. A third group of parameters influencing the lifetime of components is related to the thermal environment of a component. In practice it will not be possible to categorize every influence in one of these groups. There will remain a group of parameters, such as air pressure and humidity, described as the climatological stress parameters. It is important to know that, generally speaking, these stressors show a strong interaction. For example, the temperature of a component will be dependent on the electrical environment (dissipated power), the environmental temperature and the thermal resistance to the environment of this component. All parameters may influence the lifetime of this component. In this case it is possible to define an "electrical/thermal stressor", a "thermal resistance stressor" and an "environmental temperature stressor". In all cases the actual device temperature will be the determining factor for degradation processes. To limit the possible number of stressors the term *basic stressors* is introduced.

A basic stressor is defined as a stressor with a direct, physical   (3.6)
influence on a failure mechanism of a component. It is not (or
only partly) possible to describe basic stressors in terms of other
basic stressors.

In the case of the example above the device temperature will be used as stressor
for this failure mechanism. *The rest of this book will discuss only basic stressors.*

3.3.1. *Stressors defined as stochastic functions*

One of the problems of the use of stressors is the difficulty to describe the effects
of a wide range of signals in terms of stressors. In practical circuits the actual
stressors might vary within a wide range, not only to variations in *normal
operation* but also due to effects such as system start-up or system turn-off. One
of the ways to describe stressors in a usable way is the introduction of so-called
stochastic stressors; stressors presented in the form of stochastic signals:

A stochastic stressor, denoted as ψ_y, is a random function where   (3.7)
y characterizes the type of stressor under consideration.

See figure 3.2. Chapter 4 will describe the individual stressors and related
stochastic stressors for different part models in detail. The following sections will
outline the properties of (stochastic) stressors and will illustrate the use of
stressors by means of practical examples.

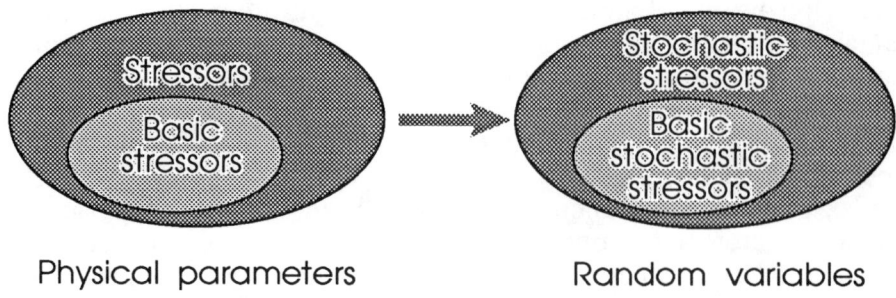

Physical parameters   Random variables

Figure 3.2: Relation stressors - stochastic stressors

3.3.2. *Stressor probability density function; single circuit, single mode*

Given a practical circuit, performing (one of) its intended function(s) under
conditions within the allowed specifications. One important group of stressors
for a certain component within this circuit is formed by electrical signals. One of

the stressors on one of the parts of such a circuit might, for example, be a certain current. A possible example of such a current is given in figure 3.3. In practice the signal forms will often be far more complicated. To compare stressors it will be necessary to introduce a comprehensive stressor characterization. For this purpose the terms *stressor probability function* and *stressor probability density function* will be introduced.

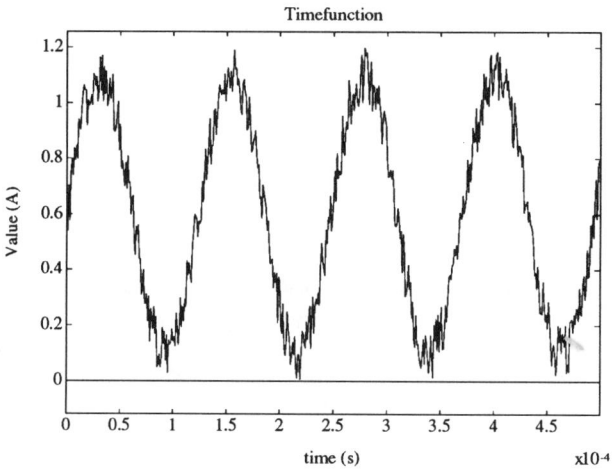

Figure 3.3: Noisy signal

The stressor probability function for entity y, $F_{str,y}(x)$ is defined (3.8)
as the probability that a stressor $\underline{\psi}_y$ does not exceed a value x.

$$F_{str,y}(x) = P(\underline{\psi}_y \leq x)$$ (3.9)

$$f_{str,y}(x) = \frac{d}{dx} F_{str,y}(x)$$ (3.10)

It is difficult to describe the probability that a stressor has exactly a certain value. Therefore a discrete probability density function will be used:

$$f_{str,y}(x) = \begin{cases} 0 & x \leq x_1 - \dfrac{\Delta x}{2} & (3.11) \\[2ex] f_{str,y}(x_1) & x_1 - \dfrac{\Delta x}{2} < x \leq x_1 + \dfrac{\Delta x}{2} \\[2ex] f_{str,y}(x_2) & x_2 - \dfrac{\Delta x}{2} < x \leq x_2 + \dfrac{\Delta x}{2} \\[1ex] \quad \vdots & \qquad \vdots \\[1ex] f_{str,y}(x_k) & x_k - \dfrac{\Delta x}{2} < x \leq x_k + \dfrac{\Delta \dot{x}}{2} \\[1ex] \quad \vdots & \\[1ex] 0 & x_n - \dfrac{\Delta x}{2} < x \end{cases}$$

where $x_i < x_{i+1}$

The corresponding discrete probability function will be:

$$F_{str,y} = \sum_{i=1}^{k} f_{str,y}(x_i) \, \Delta x \qquad x_k - \frac{\Delta x}{2} < x \leq x_k + \frac{\Delta x}{2} \tag{3.12}$$

$$F_{str,y} = \sum_{i=1}^{\infty} f_{str,y}(x_i) \, \Delta x = 1 \tag{3.13}$$

In practice the stressor probability function and the stressor probability density function of a continuous signal will be obtained by using discrete samples in time-intervals Δt. This sampling process can be described by its sample frequency $f_{sample} = 1/\Delta t$. See figure 3.4 and 3.5. For an accurate analysis of practical stressors this results in the following statement:

Accurate description of stressor probability density functions will   (3.14)
require a sampling frequency greater than two times the highest
frequency in the stressor's frequency spectrum.

$$f_{sample} \geq 2.\textit{maximum}\,(f_{signal}) \tag{3.15}$$

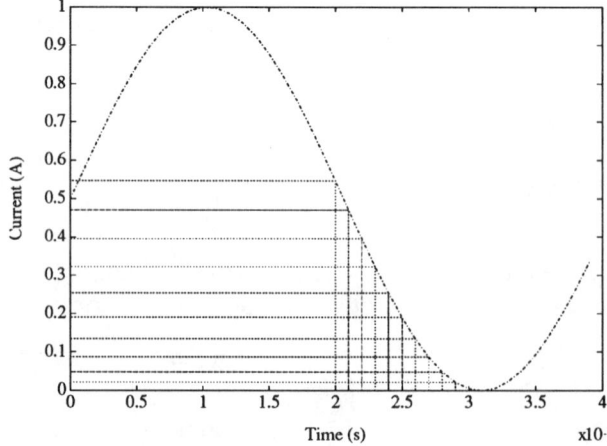

Figure 3.4: Converting signals to stressor pdf

This discrete probability density function will usually be presented in the form of a histogram. The probability density function for the time signal of figure 3.3 is given in figure 3.6.

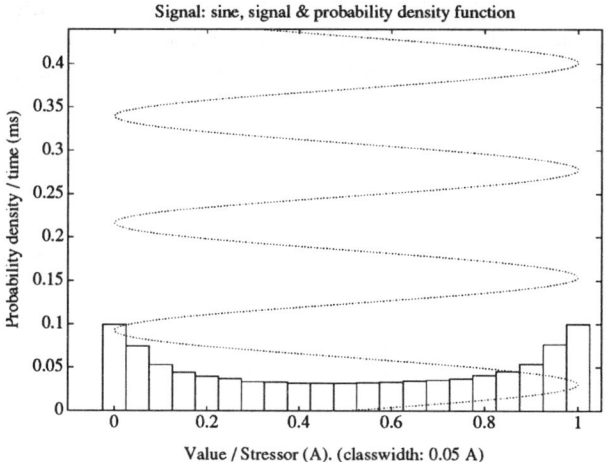

Figure 3.5: Relation stressor pdf/signal

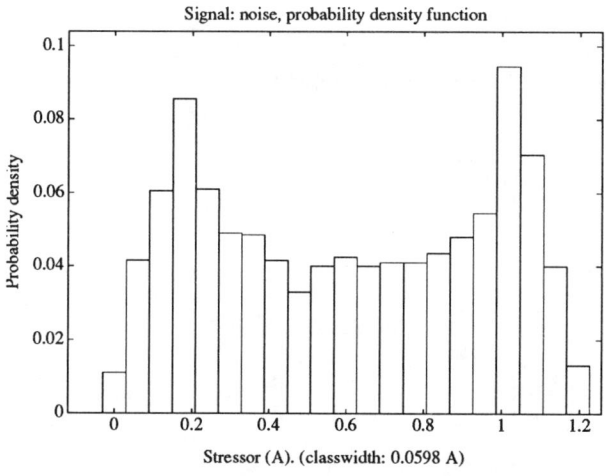

Figure 3.6: Stressor pdf of signal figure 3.3

3.3.3. *Stressor probability density function; single circuit, multiple modes*

For many applications it is not possible to derive a general applicable stressor probability density function from a single signal, related to a circuit operating in one single mode. In practical situations a circuit has more than one operation mode. The most simple case in this respect is a circuit which is switched on and off. In this case it is possible to distinguish two main stressor modes derived from the conditions *circuit on* and *circuit off*. Transition between these two operation states is quite often not a perfect abrupt switching between two system states. Quite often a circuit shows a switching behaviour more related to figure 3.7. See figure 3.8 for the corresponding stressor probability density function. Important conclusion from figure 3.8 is that the *switch-on peak* of the signal given in figure 3.7 is found in the right-hand part of the stressor probability density function.

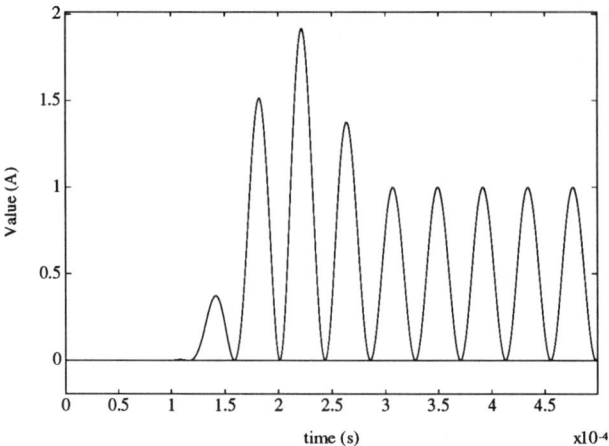

Figure 3.7: Transient switching effects (time-domain)

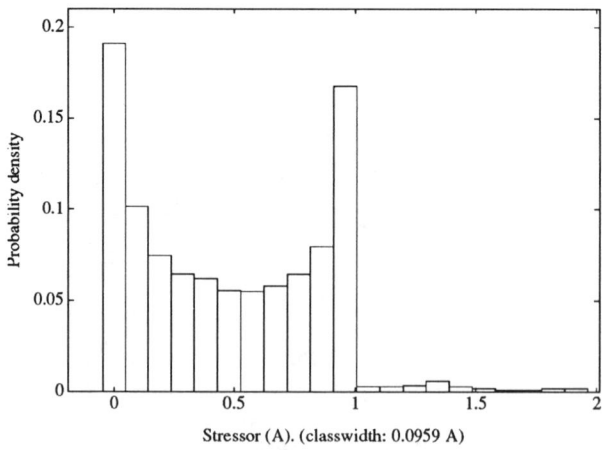

Figure 3.8: Transient effects in stressor pdf

One of the main purposes of introducing stressor probability density functions was to characterize the stress related to a certain failure mechanism by means of one single function. Therefore it is required that the stressor probability function is an ergodic description of the actual signal. For a detailed description of the required probability density function it will be necessary to have detailed knowledge about the various operating modes of a circuit and the transitions between the modes. This will result in the following statement:

> *Accurate description of stressor probability density functions will* (3.16)
> *require an observation period* $t_{total\ sample}$ *greater than the*
> *period of the lowest frequency in the stressors frequency spec-*
> *trum.*

$$t_{total\ sample} \gg maximum\ (\frac{1}{f_{stressor\ signal}})$$ (3.17)

Using this method it is possible to describe the (ergodic) probability density function for any practical stressor. However, this might result in very large numbers of samples required to describe one single stressor.

As a signal has often one or more independent quasi-stationary states, each characterized by their stressor probability density function, it is possible to derive the overall stressor probability density function from the individual state probability density functions using:

$$f_{str,y}(x) = \sum_{i=1}^{n} \frac{T_i}{T_{total}}\ f_{str,y,i}(x)$$ (3.18)

$f_{str,y,i}(x)$: the stressor probability density function of quasi-stationary state i.

$\frac{T_i}{T_{total}}$: the fraction of time that the stressor is in quasi-stationary state i

3.3.4. *Stressor probability density function; multiple circuits, multiple modes*

In practice in a large series of circuits the stressors will never be exactly the same. Due to individual differences in part parameters and environmental conditions the individual signals in circuits will show variations. As a result the individual stressor probability density functions will also show individual differences. For an example: see figure 3.10 and 3.9.

To predict the behaviour of a large series of circuits it will be necessary to introduce a new term: mean stressor probability density function:

$$\overline{f}_{str,y}(x) = \lim_{n \to \infty} \frac{\sum_{i=1}^{n} f^{i}_{str,y}(x)}{n}$$ (3.19)

$f^{i}_{str,y}(x)$: the stressor probability density function of circuit i.

It is important to distinguish a number of important differences between single stressor probability density functions and mean stressor probability density

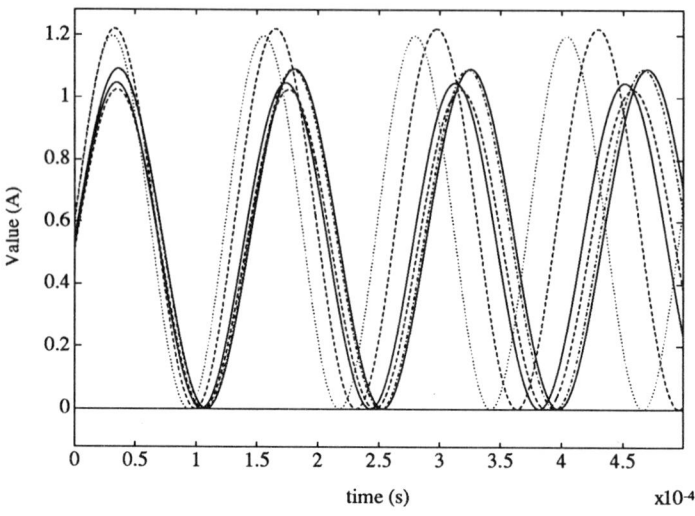

Figure 3.9

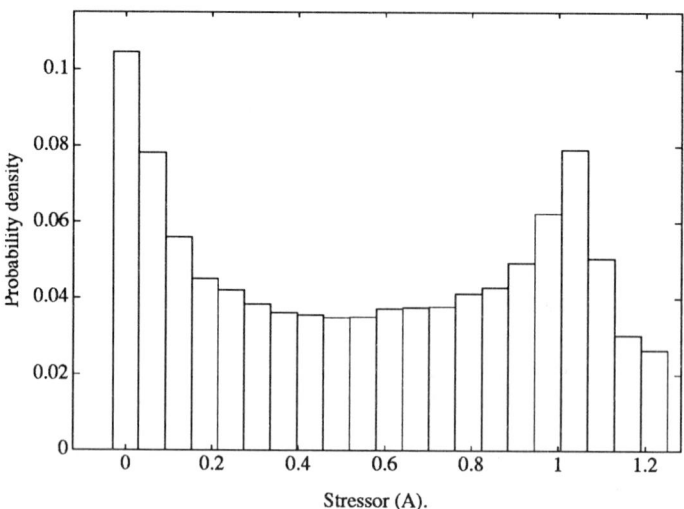

Figure 3.10: Stressor pdf batch production

functions. As these differences are very important in the calculation of part failure probabilities these differences will be discussed in the corresponding sections (See section 3.4 and figure 3.11.)

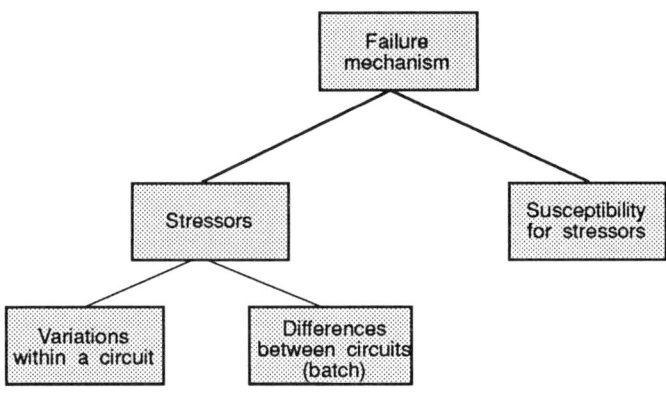

Figure 3.11: Influence factors mean stressor function

3.3.5. *Multi-variable stressor probability density functions (practical example)*

In the previous chapters a method was described to derive stressor probability density functions from practical signals. These probability density functions are related to failure mechanisms occurring in electronic components. Some failure mechanisms are influenced by more than one correlated stressor. For those failure mechanisms it will be necessary to use combined stressor probability- and probability density functions. For example, when a certain failure mech-

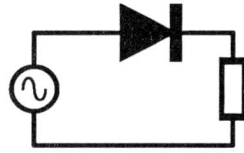

Figure 3.12

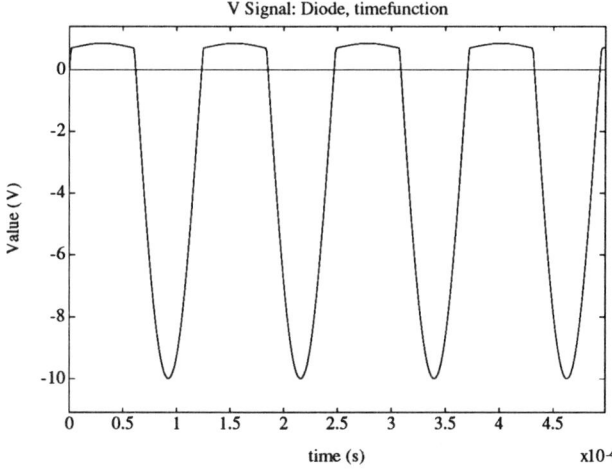

Figure 3.13: Diode voltage

anism in a diode (see Chapter 4 for more detailed examples) depends on both the momentary current and the momentary voltage it will be required to have not only the standard stressor probability (distribution) functions but also the joint voltage/current stressor probability (distribution) functions. See figures 3.12 (circuit), 3.13 and 3.14 (time-signals) and 3.15 (joint probability density function).

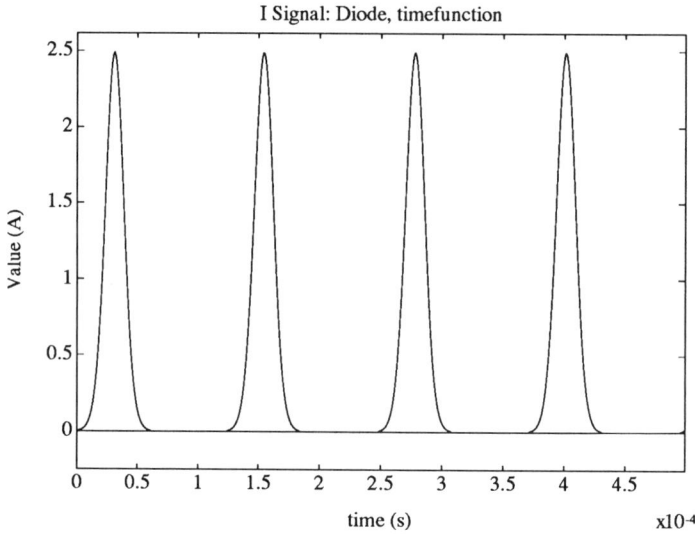

Figure 3.14: Diode current

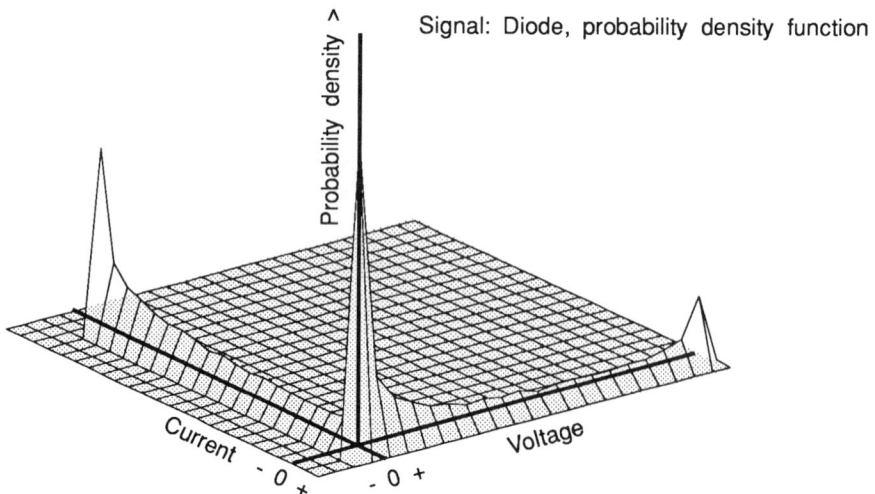

Figure 3.15: Joint stressor probability density funct.

Where the previous sections discussed stressors as abstract parameters, the following sections will give some practical examples of failure mechanisms and related stressors.

3.3.6. *Examples of practical stressors*

3.3.6.1 *Electrical stressors*

As discussed in the previous sections it is possible to distinguish different stressor groups: electrical stressors, thermal stressors and mechanical stressors. This book will mainly concentrate on failure mechanisms related to electrical (overstress) failure mechanisms. More details considering electrical stressors are discussed in Chapter 4 as a part of the description of the various part models. A short summary of some electrical stressors with related failure mechanisms is given below.

Failure mechanism	Stressor
Current breakdown	Current density
	Environmental temperature
Thermal cracks	Dissipated power
	Environmental temperature
High-voltage breakdown,	
(Impact ionization)	Electric field
(Avalanche and Zener)	Electric field
(Electron trap ionization)	Electric field
	Environmental temperature
	Time dependent behavior of the electric field: $\mathrm{d}\dfrac{E}{\mathrm{d}}t$
Corrosion (time dependent)	Environmental temperature (negative influence on susceptibility)
	Dissipated power (negative influence on susceptibility)
	D.C. voltage
	Moisture
Electromigration (time dependent)	Current density
	Environmental temperature
Secondary diffusion (time dependent)	Environmental temperature
	Power dissipation

Failure mechanism	Stressor
Switch-on effect (diodes)	Re-polarization speed $\mathrm{d}\frac{V}{\mathrm{d}}t$
	Speed of charging $\mathrm{d}\frac{I}{\mathrm{d}}t$
Switch-off effect (diodes)	Stored charge Q_s in the diode at the moment of polarity reversal
	Re-polarization speed $\mathrm{d}\frac{V}{\mathrm{d}}t$
	Maximum reverse diode current
	Applied reverse voltage
Forward bias second break-down (bipolar transistors)	Collector Emitter voltage
	Slope of the base current during switching-on $\mathrm{d}\frac{I_b}{\mathrm{d}}t$
	Slope of the collector current during switching-on $\mathrm{d}\frac{I_c}{\mathrm{d}}t$
	Environmental temperature
Reverse bias second break-down (bipolar transistors)	Collector Emitter voltage
	Stored charge in the transistor collector current at the moment of transistor switch-off
	Discharge speed $\frac{dI}{dt}$ positive influence: *fast discharging of the transistor base* negative influence: *too fast discharging of the transistor base will cause "charge bubbles".*
	Environmental temperature

The electrical stressors are for a large part determined by the electrical behavior of a circuit. Especially for the determination of stressors it is important to know not only the "normal" behaviour of a circuit but also the *marginal* conditions such as switching on and switching off.

As the electrical stressors are related to the electrical signals existing in a circuit it will be necessary to have detailed knowledge about these signals. Using the techniques discussed in the previous sections it is possible to describe most electrical stressors in terms of probability density functions.

3.3.6.2 *Thermal stressors*

Thermal effects in components are related to two main influence factors: the heat generated in the component (see also electrical stressors) and the ambient temperature. Thermal effects are very important stressors for many electronic components. (See also Chapter 4) First of all the speed of many physical (degradation) processes is related to the temperature via Arrhenius law:

$$K = A\, e^{\frac{-E_a}{kT}}$$
(3.20)

K: Temperature reaction rate

A: Constant

E_a: Activation energy of the process

k: Boltzmann constant

T: Absolute temperature

A second problem, related to temperature is the problem of thermal cycling. (See also electrical/mechanical stressors) Due to different thermal expansion coefficients of materials used in electronic components it is possible that a sudden increasing or decreasing temperature causes mechanical stresses in materials. This may cause material cracks after a certain time.

3.3.6.3 *Mechanical stressors*

Mechanical stresses, especially stresses related to thermal effects and effects of vibrations, may have a great influence on some failure mechanisms of electronic components. Especially for components with a comparatively large mass, vibration might be one of the most important failure causes.

Other mechanical aspects such as the mechanical stress on components may also have influence on failure probabilities but, as those aspects are more related to the mechanical structure of circuits, these aspects are not discussed in detail here.

3.4. Susceptibility for (combinations of-) stressors

In the previous sections stressors were derived as stochastic signals, influencing the lifetime of a component. To predict the lifetime of a component it is not only necessary to know the behaviour of stressors, often derived from the functional aspects of a circuit, but also to know the effect of these stressors on the lifetime of such a component. Generally speaking the persistence against a certain failure mechanism will be dependent on the time and often also on the history of a component. To describe the effects of stressors on the life-time of a circuit the term *susceptibility* is introduced.

The susceptibility of a component to a certain failure mechanism   (3.21)
is defined as the probability function indicating the probability
that a component will not remain operational for a certain time
under a given combination of stressors. The susceptibility re-
lated to a failure mechanism y will be denoted as
$S_y(t,\psi_p,\psi_q,\psi_r)$

In accordance with the previously described stressor probability- and stressor probability density function it is also possible to define a susceptibility density function defined as:

$$s_{y,p}(\psi_p) = \frac{d}{d\psi_p} S_y(\psi_p)$$ (3.22)

3.4.1. *One variable catastrophic susceptibility model*

The most simple susceptibility model is the *one- variable, time- independent catastrophic susceptibility model*. This model uses the following relation:

$$S_y(\psi) = \begin{cases} 0 & \psi < \psi_0 \\ 1 & \psi \geq \psi_0 \end{cases}$$ (3.23)

$$s_y(\psi) = \begin{cases} 0 & \psi < \psi_0 \\ \delta & \psi = \psi_0 \\ 0 & \psi > \psi_0 \end{cases}$$ (3.24)

See also figure 3.16 and 3.17. This means that a component fails immediately at the moment a certain stressor exceeds a certain threshold value. On the other hand the component will remain unchanged in case of stressors below this threshold value. As the model is assumed to be time independent this threshold

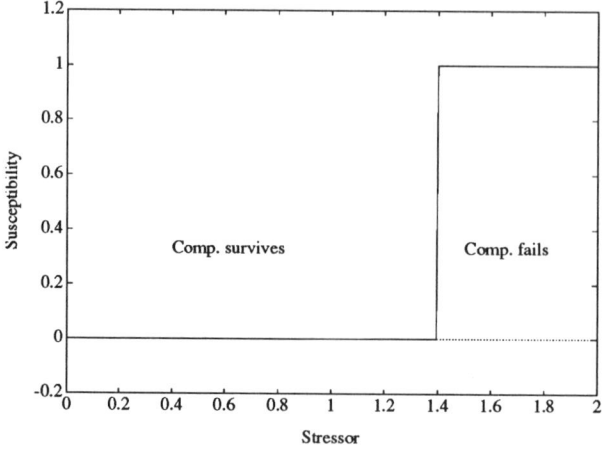

Figure 3.16: One-variable catastrophic susceptibility

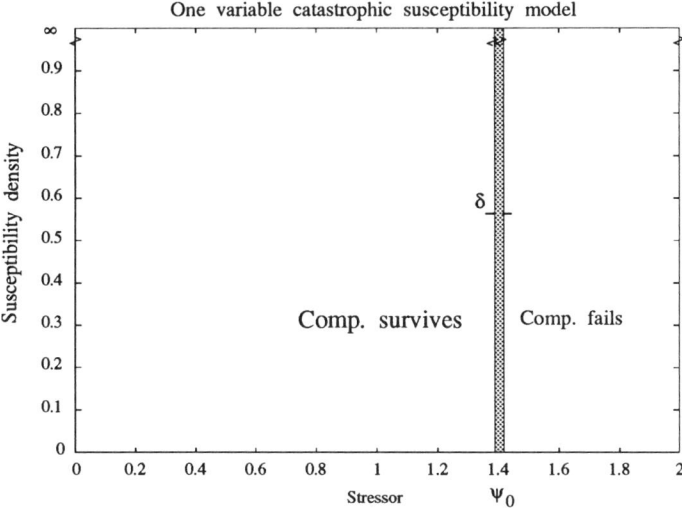

Figure 3.17: One-variable catastrophic susc. pdf

value will not change with time. Therefore the model is memory-less. Stressors in the past have either destroyed the component by exceeding the threshold value or left the component unchanged by remaining under this threshold value.

3.4.2. *Multi- variable catastrophic susceptibility models*

In practice the previous model is, in many cases, too simple. Components will often fail due to a combination of stressors. In this case it is not possible to speak of threshold stressors. This can be illustrated by the following sample model:

$$S_y(\psi_p, \psi_q) = \begin{cases} 0 & \psi_p < c_1 - c_2 \ \psi_q \\ 1 & \psi_p \geq c_1 - c_2 \ \psi_q \end{cases} \tag{3.25}$$

$$s_y(\psi_p, \psi_q) = \begin{cases} 0 & \psi_p < c_1 - c_2 \ \psi_q \\ \delta & \psi_p = c_1 - c_2 \ \psi_q \\ 0 & \psi_p > c_1 - c_2 \ \psi_q \end{cases} \tag{3.26}$$

See also figure 3.18 and 3.19. In this case the "threshold" for one stressor is dependent of another stressor. A simple example of a multi- variable suscepti- bility model is found in the well-known susceptibility of a resistor for power-over- load. The related failure mechanisms depend on two stressors; the current through the resistor and the environmental temperature. It is not possible to distinguish thresholds for either stressors, as the burn out of the resistor will

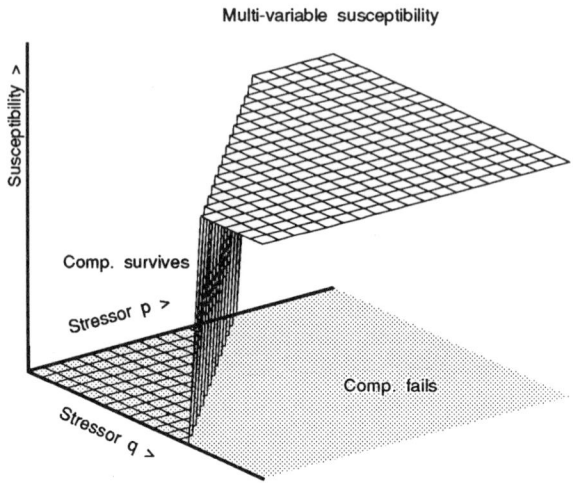

Figure 3.18: Two variable catastrophic susceptibility

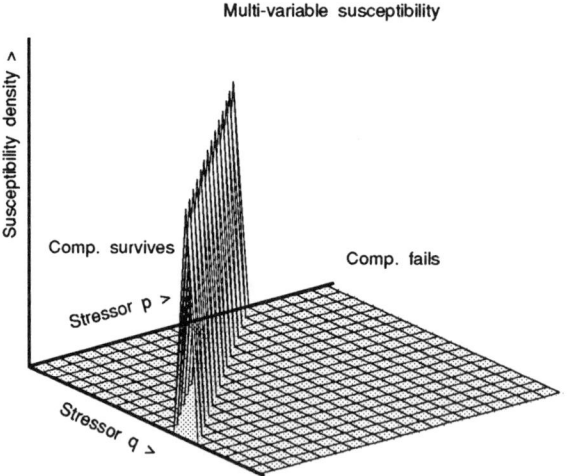

Figure 3.19: Multi-variable susceptibility pdf

depend on the existing combination of both stressors . Detailed examples of such multi-variable models will be discussed in Chapter 4.

3.4.3. *Gradual susceptibility models*

One of the main problems of susceptibility models is the certainty that in practice the actual number of stressors influencing a failure mechanism will be larger than the number of stressors described in a susceptibility model. Parameters like air-humidity, mechanical stability, etc. may have influence on failure mechanisms. The first problem is that it will be quite difficult to find stressor distributions of these parameters; the second problem will be the difficulty to model these stressors in a susceptibility model. One of the solutions of taking these stressors into account is *incorporating* these stressors as gradual effects in

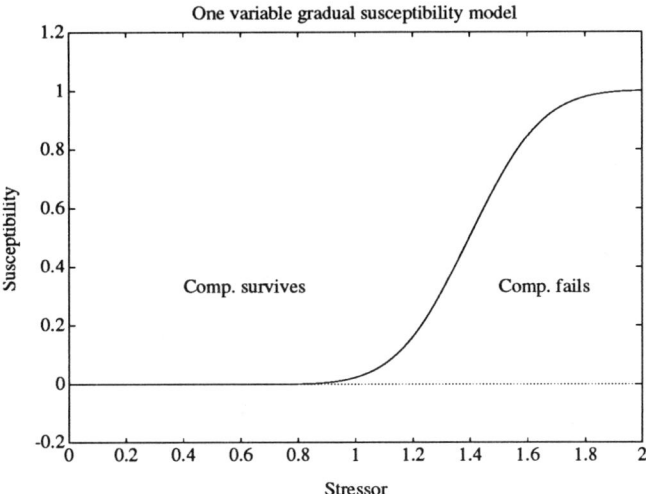

Figure 3.20: Gradual susceptibility

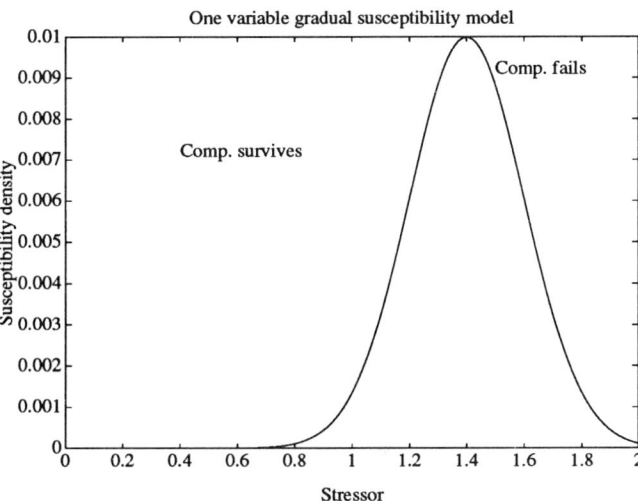

Figure 3.21: Gradual susceptibiliy density

standard susceptibility models. This will result in a susceptibility model which is not catastrophic. Due to the already mentioned *unmodellable* stressors the catastrophic susceptibility model will turn into the gradual model described in figures 3.20 and 3.21.

3.4.4. *Constantly degrading susceptibility models*

Due to a wide range of reasons (see also Chapter 4) it is possible that devices show ageing. In this context the following definition is used:

Ageing is defined as the (with time) increasing susceptibility of   (3.27)
a device for its failure mechanisms.

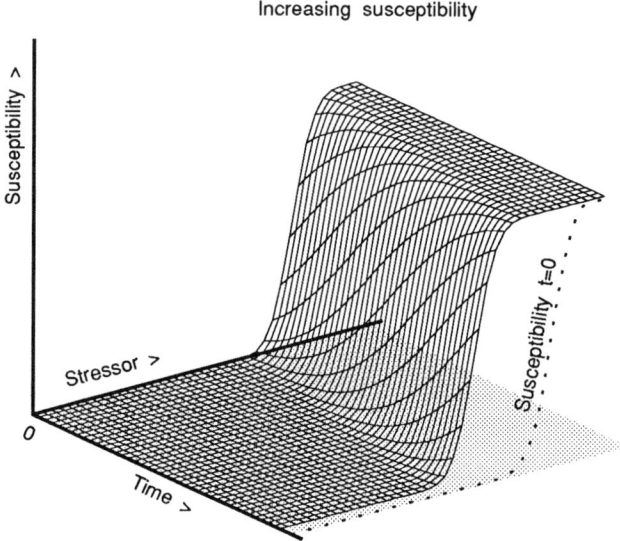

Figure 3.22: Degradation (increasing susceptibility)

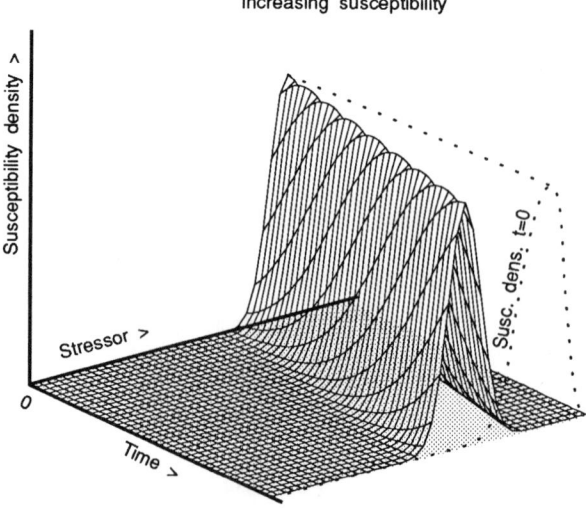

Figure 3.23: Degradation, susceptibility pdf

A simple ageing model is presented in figures 3.22 and 3.23. This model describes a linear increasing susceptibility model. It is possible to apply this model on failure mechanisms where failure mechanisms with independent physical degradation mechanisms are involved. In many cases the susceptibility is not linear increasing. In many cases the *history* of a component will play an important role in a components susceptibility. The term history is used to describe the influence of stressors in the past on the present susceptibility of a component for a certain failure mechanism.

3.4.5. *Susceptibility models for large series components*

The problem with large series of components is that not even two components within such a series will have the same susceptibility. The susceptibility is strongly dependent on the physical structure of a component. Due to variations in the production process of components every component will have its own susceptibility. To describe the susceptibility of a large series of components the term *mean-susceptibility* is introduced. See also figure 3.24.

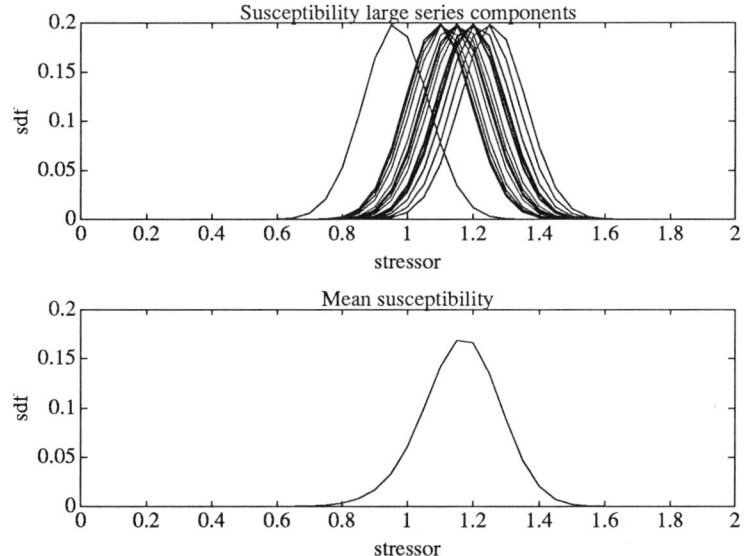

Figure 3.24: Mean susceptibility density batch

$$\overline{S}_y\,(x) = \lim_{n \to \infty} \frac{\displaystyle\sum_{i=1}^{n} S_y\,(x)}{n} \qquad (3.28)$$

$$\bar{s}_y(x) = \lim_{n \to \infty} \frac{\sum_{i=1}^{n} s_y(x)}{n}$$

(3.29)

3.4.6. *Weak sub-populations*

In many cases a large series of components does not show a susceptibility as described in the previous section. Quite often devices exist with certain built-in weaknesses. Examples are devices with poor bondings and devices with material pollution. It is possible to model such a series of components using a bi-modal distribution. One part of such a distribution meets with a distribution with standard mean-susceptibility, the other part of the distribution has a much higher mean-susceptibility. See figure 3.25.

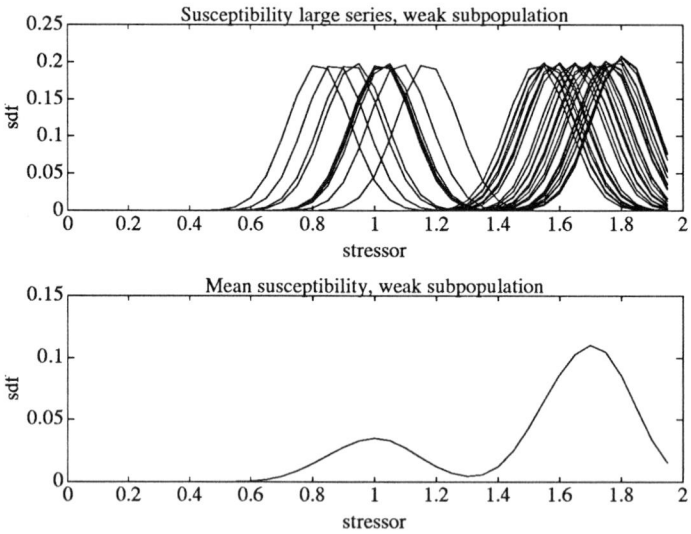

Figure 3.25: Weak subpopulation susceptibility

3.5. **Failure probability and reliability**

Although the susceptibility of a part is related to failure mechanisms within a part the susceptibility alone is not usable to describe or predict part failures. Part failures are related to a combination of stressors and susceptibility. In this paragraph stressors and susceptibility of failure mechanisms are combined to calculate part failure probabilities. See figure 3.26.

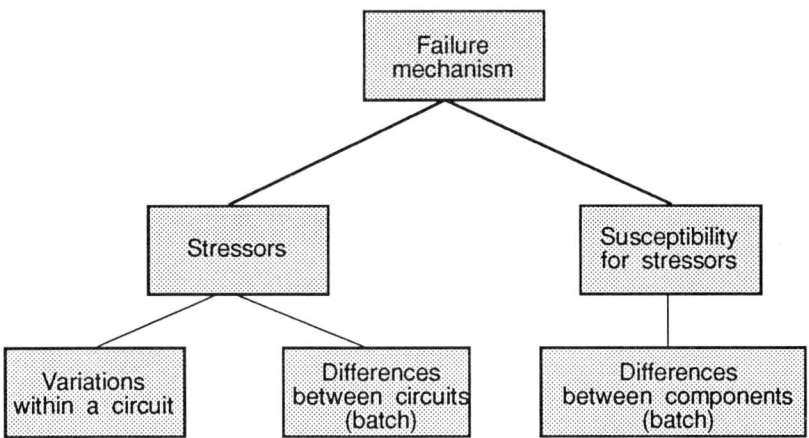

Figure 3.26: Stressor/susceptibility influences

3.5.1. *Failure probability for single failure mechanisms*

As stated in the previous section analysis of failure probabilities requires detailed analysis of both stressors and susceptibility. Given a part with one single failure mechanism influenced by one single stressor. Stressor and susceptibility density are given in figure 3.27. It is obvious that the main source of problems is the overlap between stressor and susceptibility density.

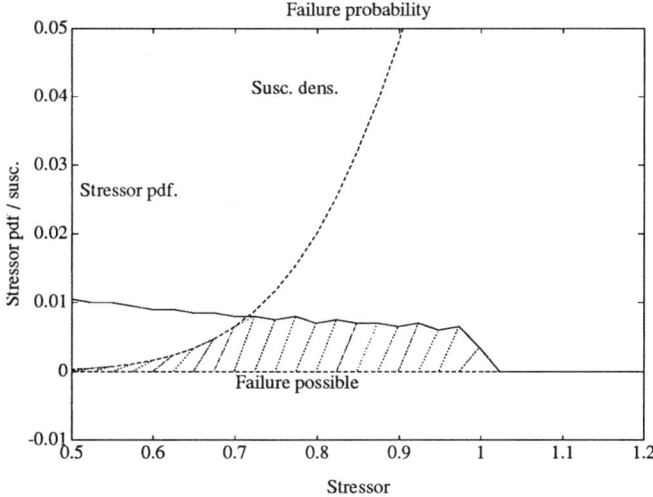

Figure 3.27: Stressor/susceptibility interaction

The first step is to calculate the failure probability for this stressor distribution on a failure mechanism with a single, one variable, time- independent catastrophic susceptibility model (such as given in figure 3.17). See figure 3.28. Failures are caused by stressors in the interval $\psi_0 \rightarrow \infty$. This results in the following failure probability:

$$f_{ail,y,\psi}(\psi_0) = \int_{\psi_0}^{\infty} f_y(\psi)\, d\psi \tag{3.30}$$

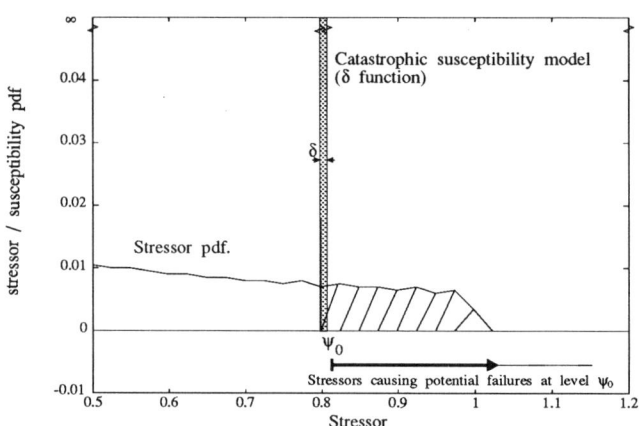

Figure 3.28: Failures due to catast. susceptibility

To calculate the failure probability also as a function of more complex susceptibility models it will be necessary to calculate the failure probability of a part of the susceptibility model, for a certain stressor interval Δ, characterized by its

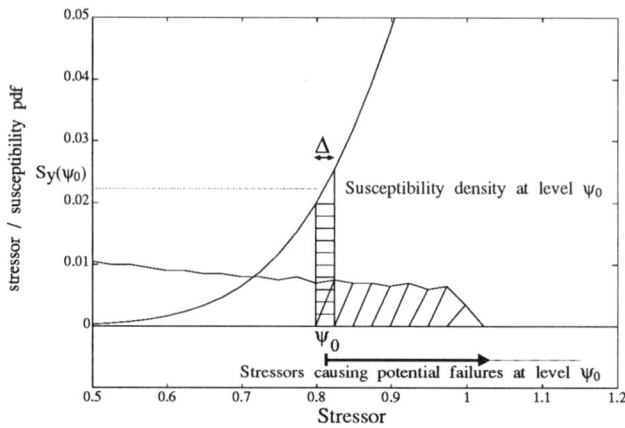

Figure 3.29: Failures due to stressor interval

mean value ψ_0 and the corresponding susceptibility density function at that point $s_y(\psi_0)$. Failures in this interval are caused by stressors in the interval $\psi_0 \rightarrow \infty$. See figure 3.29. This results in the following failure probability for this part of the susceptibility curve:

$$f_{fail,y,\psi}(\psi_0) = \Delta\,(s_y(\psi_0) \int_{\psi_0}^{\infty} f_y(\psi)\,d\psi) \tag{3.31}$$

Of course there is also the probability that the part has failed at a lower susceptibility level. This results in the possibility to predict the failure probability per time-interval of a certain failure at stressor level $\Delta\psi_0$ using the following relation:

$$f_{fail,y,\psi}(\psi_0) = \Delta\,(s_y(\psi_0) \int_{\psi_0}^{\infty} f_y(\psi)\,d\psi)\,(1-\int_{0}^{\psi_0-\Delta} f_{fail,y}(\psi)d\psi) \tag{3.32}$$

The last term is introduced to subtract failures caused by stressors at a lower susceptibility level. As, most often, failure probabilities are very small, it will be possible for many cases to simplify the previous expression to:

$$f_{fail,y,\psi}(\psi_0) = \Delta\,(s_y(\psi_0) \int_{\psi_0}^{\infty} f_y(\psi)\,d\psi) \tag{3.33}$$

if
$$(1-\int_{0}^{\psi_0-\Delta} f_{fail,y}(\psi)d\psi) \approx 1$$

See figure 3.30 for the resulting failure probability curve.

As the susceptibility is defined as the probability that a component will not remain operational during a certain *time* it is also possible to calculate the failure probability during a certain observation time t_{obs}:

$$f_{fail,y,\psi}(\psi_0) = t_{obs}*\Delta*(s_y(\psi_0) \int_{\psi_0}^{\infty} f_y(\psi)\,d\psi) \tag{3.34}$$

Important requirements for the use of the previous formula are:

The observation time t_{obs} should be larger than the total elapsed sampling time required to obtain an ergodic description of the associated stressors $t_{total\,sample}$. $\tag{3.35}$

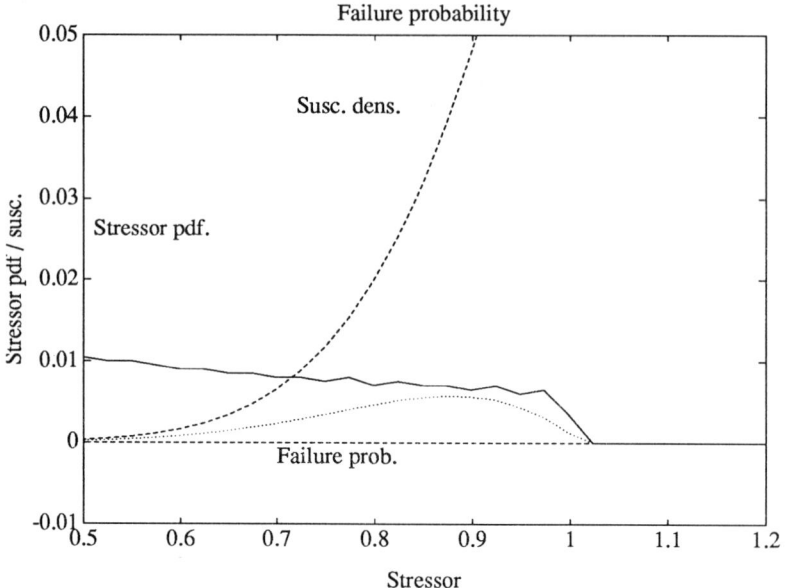

Figure 3.30: Failure probability due to interaction

$$t_{obs} > t_{total\ sample} \tag{3.36}$$

$f_{fail,y,\psi}(t,\psi)$ *is assumed to be constant during the time-interval* t_{obs}. (3.37)

From the previous equation it is possible to calculate the failure probability of a part per failmechanism per time-interval. This can be done using:

$$f_{fail,y} = \int_{0}^{\infty} f_{fail,y,\psi}\ (t,\psi)d\psi \tag{3.38}$$

The previous equation can be used to calculate the part failure probability per time-interval:

$$f_{fail} = \sum_{i=1}^{n} f_{fail,i} \tag{3.39}$$

Using the previous assumptions it is possible to calculate the probability that a component survives from t to t+dt. For this purpose (a somewhat adapted form of) the well-known reliability function is used [Sho68].

$$R(t..t+\Delta T) = \frac{\sum\limits_{i=1}^{k} \text{Devices operational at } (t+\Delta t)}{\sum\limits_{i=1}^{n} \text{Devices operational at } (t)} = R(t) \, F(\Delta t) = R(t) \, \Delta t \, f(t) \qquad (3.40)$$

Using this formula it is possible to calculate the failure probability for one single failure mechanism within one single device. The problem with such a calculation is to calculate the reliability of a component at time t. In practice a component may have more than one failure mechanism. Another problem is that within a large series of "identical" components many individual differences may exist. The next paragraphs will define a method to calculate the reliability and failure probability for components, based on stressors and susceptibilities of failure mechanisms.

3.5.2. *Component failure probability for multiple failure mechanisms*

Failure probability depends on two different entities: failure mechanisms and the stressors influencing these failure mechanisms. Generally speaking the stress on a component will vary with time, depending on a large number of stressors. The survival probability of a component will be dependent of a large number of parameters such as physical component structure, environmental conditions and the history of the component. Another problem is that most components tend to have more than one failure mechanism, resulting in more than one "reliability". Generally speaking there will be a number of "reliabilities", each related to their own physical failure mechanism.

In association with the terms failure probability and reliability, defined on the level of failure mechanisms, it is also possible to determine the failure probability and reliability on component level.

In the previous section failure mechanisms are assumed as independent entities. In many cases, however, there is a large correlation between the various failure mechanisms existing within a component. To analyse failure probability and reliability on component level the following assumptions are made:

> *The reliability of a component is assumed to be dependent of the momentary values of the stressors and the momentary physical state of a component only.* (3.41)

> *The momentary physical state of a component depends on the physical structure of a component and the behaviour of the stressors in the past.* (3.42)

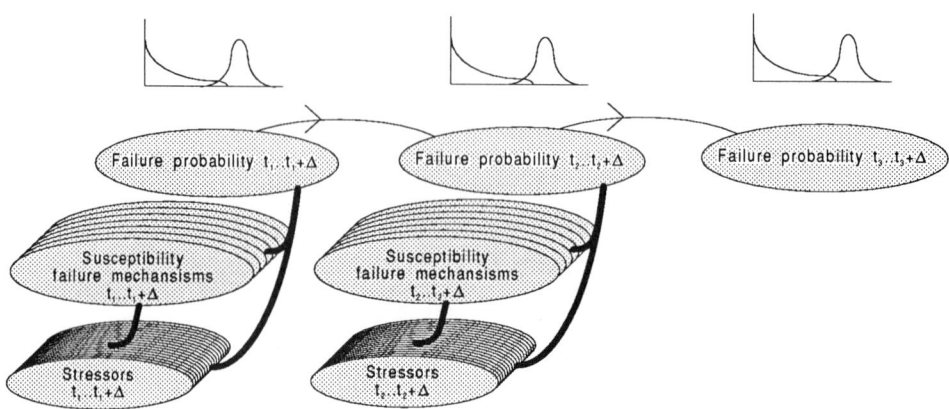

Figure 3.31: Markov processes

As, for a large series of components, the physical structures of the individual components will be different for every component, the survival probability of such a series of components will also show individual differences.

Generally speaking the stress on a component may vary with time due to circuit behaviour and circuit use. The circuit behaviour will differ amongst a series of circuits due to physical differences in the individual circuit components, the physical structure of a circuit, the use of a circuit and the environment (electrical, thermal, etc.) of the circuit. To summarize this variety of effects it is useful to describe stressors as stochastic signals with properties depending on the influence factors, mentioned above:

All stressors can be described as stochastic stressors.   (3.43)

These assumptions make it possible to derive the failure probability and reliability of a component using a Markov approach. See figure 3.31.

For a Markov approach the following requirements should be fulfilled:

Susceptibilities of all failure mechanisms in a component are   (3.44)
known and are constant in the time interval $t \rightarrow t+\Delta t$.

All stressors $\psi_a(t)$, $\psi_b(t)$,... are known as stochastic signals for   (3.45)
the time interval $t \rightarrow t+\Delta t$.

The failure probability (or reliability) should be given for a certain   (3.46)
(initial) time t.

Using these properties it is possible to calculate the reliability and failure probability for components, derived from internal failure mechanisms for time $t+\Delta t$. For this purpose the following relations are used:

$$\vec{P}(t+\Delta t) = \vec{P}(t) \, \Theta \, (\Delta t)$$ (3.47)

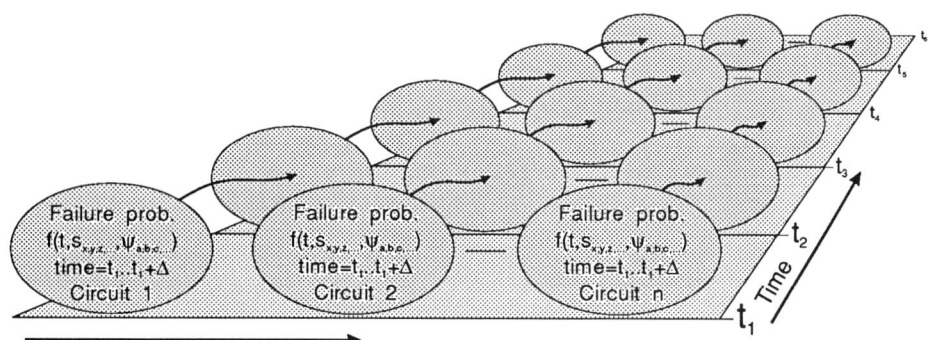

Differences between circuits
- tolerances in function (stressors)
- tolerances in susceptibility

Figure 3.32: Multiple Markov analysis to simulate batch

$$= \overrightarrow{P}(t) \begin{bmatrix} P_{x \to x} & .. & .. & P_{x \to y} \\ .. & & & .. \\ .. & & & .. \\ P_{y \to x} & .. & .. & P_{y \to y} \end{bmatrix}$$

where $\overrightarrow{P}(t)$ is the *state probability vector* of a component. This state probability vector is defined as:

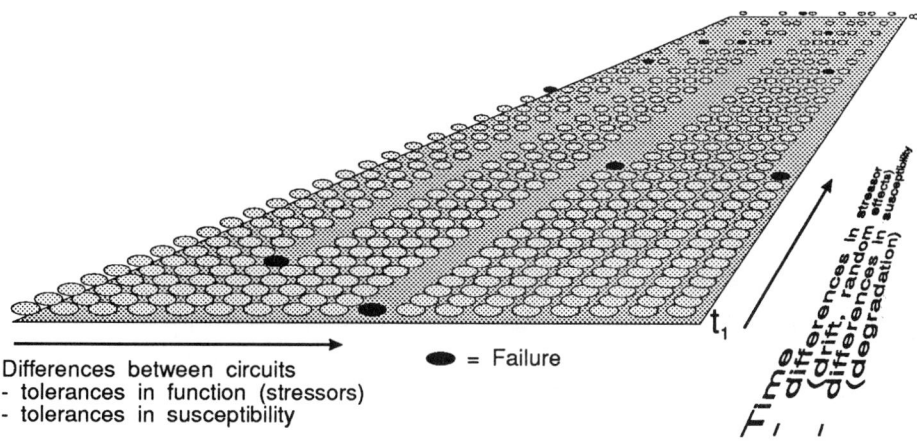

Differences between circuits
- tolerances in function (stressors)
- tolerances in susceptibility

● = Failure

Figure 3.33: Time-dependent stressor susc. analysis

$$\vec{P}(t) = \begin{cases} P_{\text{operational}}(t) & : \textit{Prob. part is operational at time t} \\ P_a(t) & : \textit{Prob. part fails due fail. mech. a at time t} \\ P_b(t) & : \textit{Prob. part fails due fail. mech. b at time t} \\ .. & \quad .. \\ P_n(t) & : \textit{Prob. part fails due fail. mech. n at time t} \end{cases} \tag{3.48}$$

$$\sum_{j=1}^{n} P_j(t) = 1 \tag{3.49}$$

$$P_1(t) = P_{\text{operational}}(t) = R(t) \tag{3.50}$$

$$P_{2..n}(t) = P_{\text{fail},2..n}(t) = F_{\text{fail},2..n}(t) \tag{3.51}$$

$$P_{x \to y} = P(y(t+\Delta t) \mid x(t)) = f_y(t)\,\Delta t\,P_x(t) \tag{3.52}$$

It is possible to replicate this calculation process for a whole batch of circuits. See figure 3.32. In this case for every circuit the individual stressor/susceptibility interaction is calculated thus simulating batch behaviour. Using this method it is possible to derive the failure probability for many parts in many practical situations, aslo in cases that considerable differences (in stressors and susceptibility) exist within a batch. See figure 3.33. This will be illustrated in the following sections.

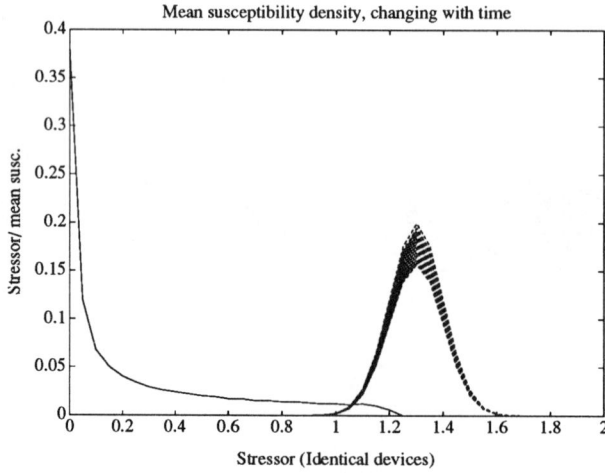

Figure 3.34: Identical devices; population shrinks

3.5.3. *Components with identical constant susceptibility*

Although identical components and identical circuits in practice do not exist, the

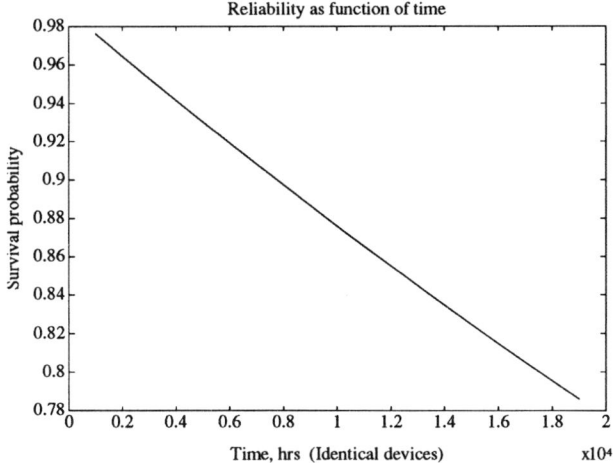

Figure 3.35: Reliability, identical devices

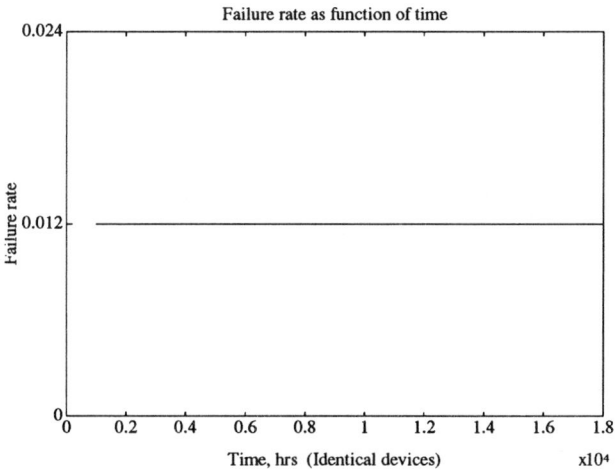

Figure 3.36: Relative failure rate

following example will be used to illustrate the consequences of the use of such a series of components. Given a large series identical components, used in a large series of identical circuits. Such a series will not only have identical functional behaviour but also identical susceptibility.

To calculate the long-term effects of the interaction between stressors and susceptibility for a large series of components it is useful to introduce the susceptibility of a failed device.

For calculation reasons the susceptibility of a failed device is   (3.53)
assumed to be a constant function at level 0.

Sfailed device$(\psi) = 0$ (3.54)

As susceptibility and stressors for all devices are exactly the same it is possible
to calculate the failure probability per time interval as a constant. This means
that a constant fraction per time interval of the remaining population is failing. It
is easy to derive that, using 3.51 for this combination of stressors and susceptibility the following relation can be derived:

$$R\,(\,t + \Delta t\,) = (\,1 - K\,\Delta t\,)\,R(t) \;\Rightarrow\; \frac{R\,(\,t + \Delta t\,) - R(t)}{\Delta t} = -\,K\,R(t) \qquad (3.55)$$

$$R(t) = e^{-K.t} \qquad\qquad\qquad\qquad\qquad\qquad\qquad\qquad\qquad (3.56)$$

Figure 3.34 shows the mean susceptibility of this series of components, changing with time. Figure 3.35 shows the calculated reliability and figure 3.36 the
resulting *relative failure rate* or *failure rate of the remaining population*. This
relative failure rate is defined as:

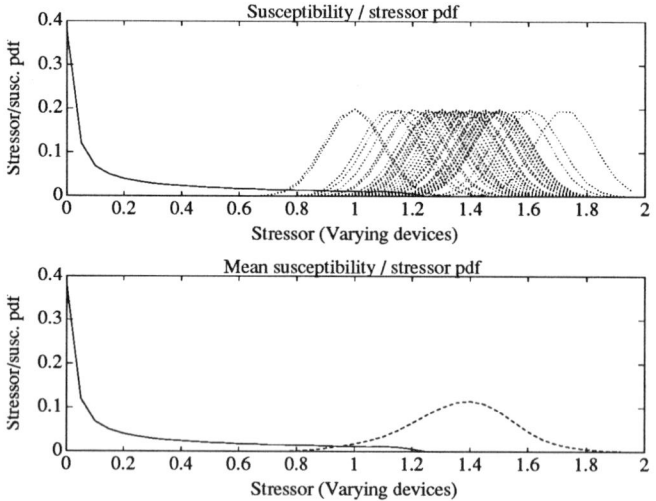

Figure 3.37: Different but constant susceptibility

$$\lambda(t) = \frac{f(t)}{R(t)} \qquad\qquad\qquad\qquad\qquad\qquad\qquad\qquad\qquad (3.57)$$

This function is often called *the hazard rate* in the literature but the same term
hazard rate is also used in the literature to calculate the probability of an unsafe
situation in high-safety applications. Therefore in this publication the term

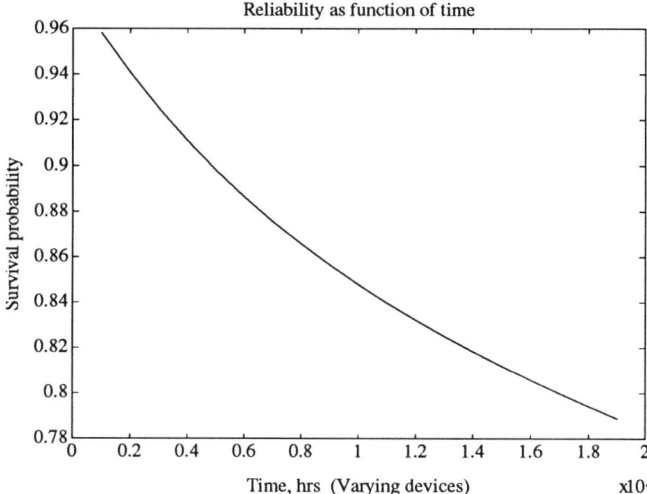

Figure 3.39: Reliability different but constant susc.

relative failure rate will be used. In case of constant identical susceptibilities it possible to derive from (3.57) and (3.58):

$$\lambda(t) = \lambda_{\text{constant}} \tag{3.58}$$

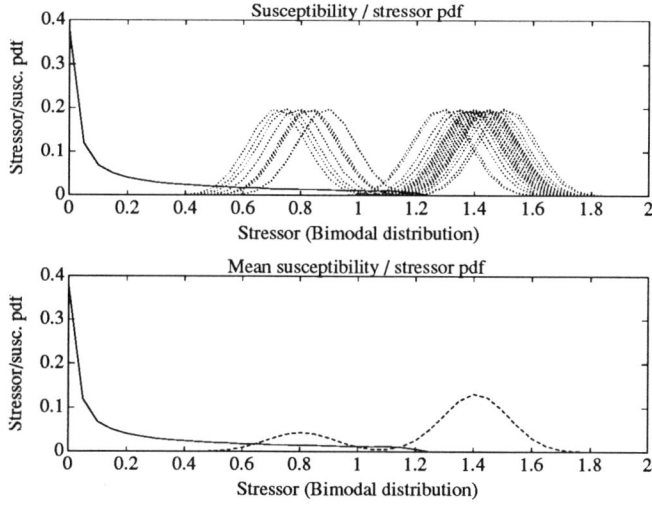

Figure 3.38: Initial distribution weak sub-population

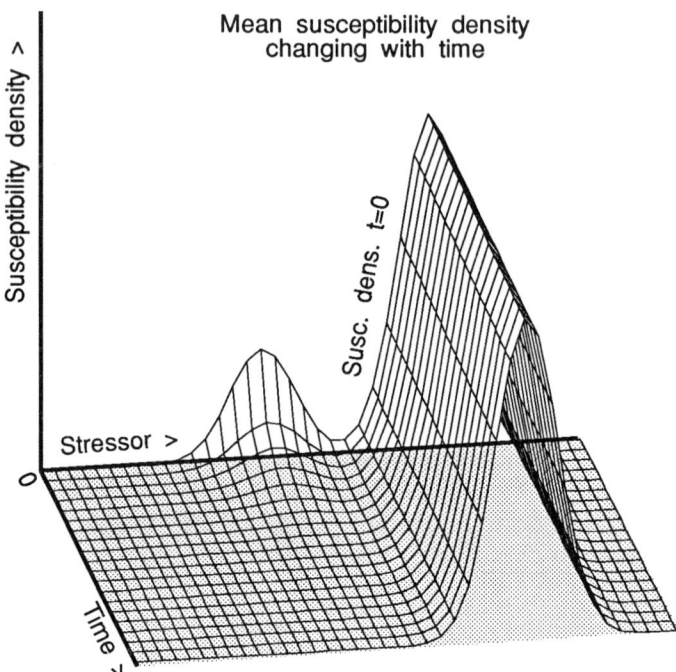

Figure 3.40: Changing susceptibility distribution

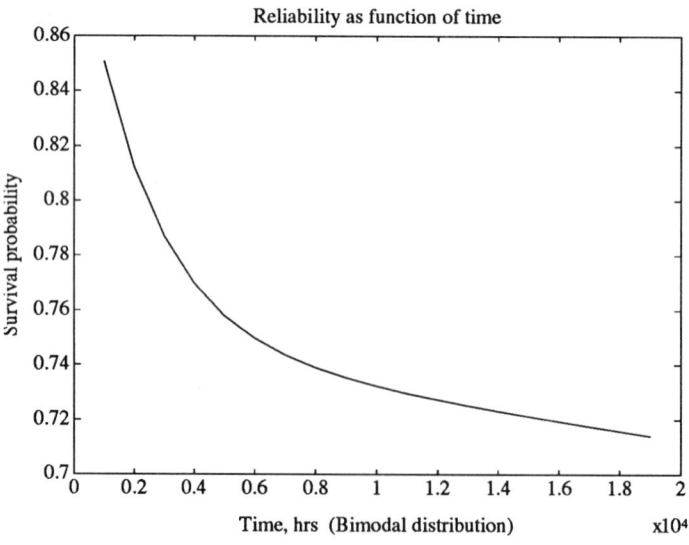

Figure 3.41: Reliability function weak sub-population

3.5.4. *Components with different but constant susceptibility*

In many cases the susceptibility of components for certain failure mechanisms will be different for every component. The differences will vary from minor differences due to instabilities in the production process of components to large differences due to (often unwanted) changes in the production process of these components. These changes will result not only in changes in the susceptibility of components but also in changes in functional behaviour. Due to these factors

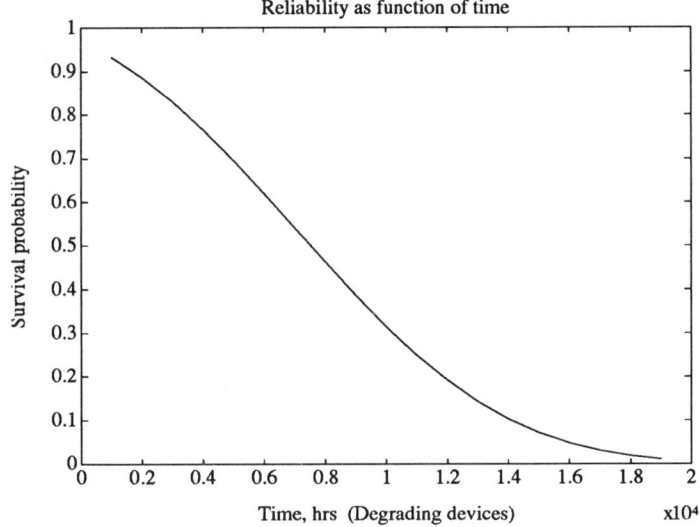

Figure 3.42: Reliability effects of degradation

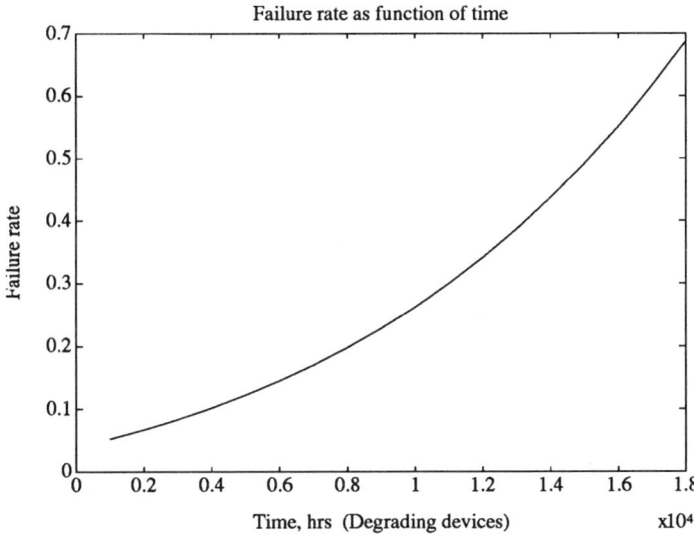

Figure 3.43: Increasing failure rate; degradation

it is important that formulae (3.32) to (3.40) are applied first on the level of individual components. From the results of these calculations it is possible to derive mean values for large series components:

> *For a detailed calculation of reliability for a series of non-*   (3.59)
> *identical components it is necessary to combine stressors and*
> *susceptibility on the level of single parts.*

The initial population of a component series with different constant susceptibility is given in figure 3.37. (individual susceptibility and mean susceptibility). Reliability is given in figure 3.39.

3.5.5. *Weak sub-populations*

In many cases a large series of components does not show a normal susceptibility distribution. In many cases devices exist with certain built-in weaknesses. Examples are devices with poor bondings, devices with material pollution, etc. It is possible to model such a series of components using the bi-modal distribution presented in section 3.5.6. One part of such a distribution meets with a distribution with standard mean susceptibility, the other part of the distribution has a much lower mean susceptibility.

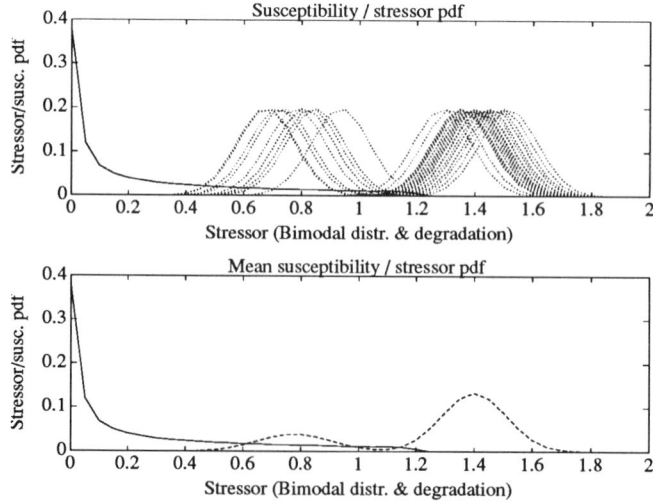

Figure 3.44: Multiple effects, initial distributions

Figure 3.38 shows the initial distribution, figure 3.40 shows the time dependent susceptibility density distribution. Figure 3.41 shows the reliability function.

3.5.6. *Degradation effects*

One of the problems that may exist in electronic parts is the problem of degradation. In the previous paragraphs the reliability of components is as-

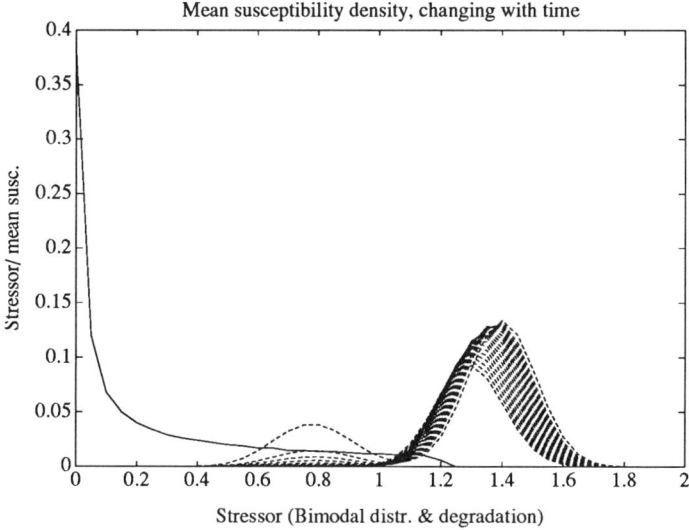

Figure 3.45: Multiple effects, time dependent

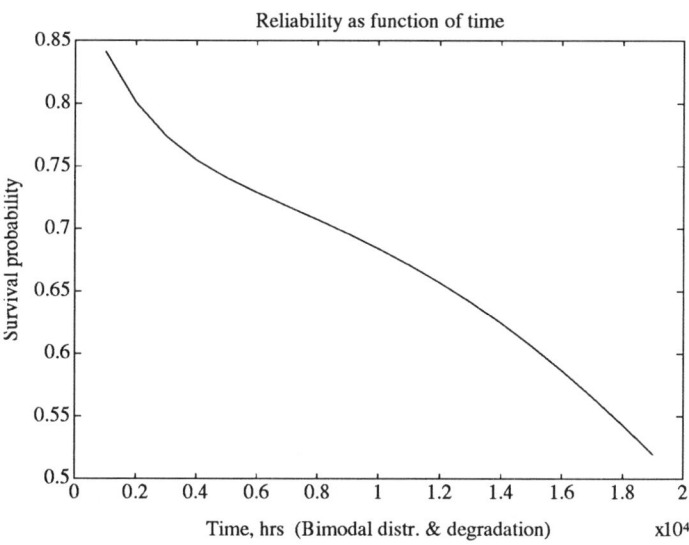

Figure 3.46: Reliability, multiple effects

sumed to be time-independent. Due to a number of effects it is also possible that the reliability of components decreases with time. Especially physical degradation processes may have influence on the reliability of certain components. The example below shows the effects of a linear decreasing susceptibility density function. For practical components the actual distribution will strongly depend

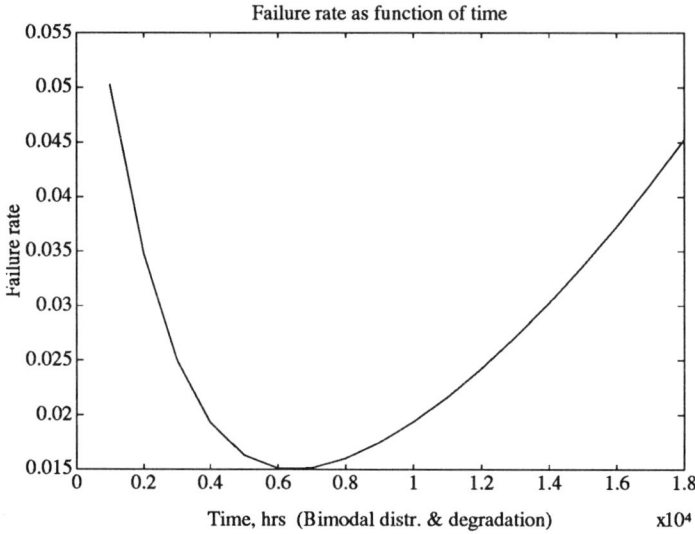

Figure 3.47: Failure rate, multiple mechanisms

on the physical structure of the component. Also the actual degradation speed will depend on this structure.

Figures 3.42 and 3.43 show the reliability and relative failure rate function as effect of a constant stressor. Interesting effect is the increasing failure rate as effect of increasing susceptibility. This effect appears to be closely related to the wearout phenomena marking the wear-out period of the traditional bathtub (see also figure 2.3) failure rate curve.

3.5.7. *Gradual failure mechanisms, cumulative effects*

The problem with multiple gradual failure mechanisms is the experience that they tend to show cumulative effects. As the susceptibility of a device is strongly related to the current physical (degradation) state of a device, another problem is the susceptibility of a device for a certain failure mechanism. This susceptibility is correlated to the susceptibility of other failure mechanisms in the same device. It is possible to model these effects using the presented methodology. The result will often be failure mechanisms with non-linear decreasing susceptibility. See chapter 4 for practical examples.

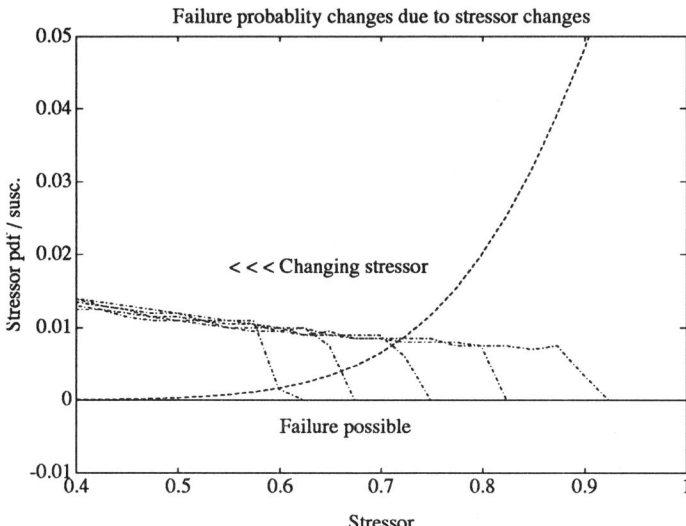

Figure 3.48: Reliability sensitive to changing stressor

3.5.8. *Combined effects*

In practical situations components will have more than one failure mechanism. The previous sections served mainly the purpose of proving the possibility to derive component failure probabilities and related relative failure rate functions from a mathematical combination of stressors and failure mechanisms. Figures 3.44, 3.45, 3.46 and 3.47 show an example of a practical component with a more complicated susceptibility. This example uses a bimodal initial distribution combined with degradation effects. An important conclusion from figure 3.45 is the possibility to derive the different susceptibility functions from such a curve. A steeply decreasing failure probability density function indicates weak sub-population(s) while the tail in figure 3.47 indicates degradation effects. This so-called bathtub curve is a well-known relative failure rate function for a wide range of electronic components. In the past this curve was not derived from theoretical models but from practical observations.

3.6. **Failure probabilities in terms of design parameters**

In the previous sections some examples of reliability curves were discussed. The problem with a practical situation is that for most devices more than one failure mechanism exists, each mechanism with several stressors. A detailed study of all possible fail-mechanisms in a wide range of electronic parts is, in it self, a very useful (and tedious) job. Purpose of this project is providing the design team of a certain system or circuits with a methodology to optimize especially the reliability aspects of such a system or circuit.

One of the main purposes of optimizing a circuit is removing possible problem sources. In terms of stressors and susceptibilities this means optimizing stressor/susceptibility relations in case of critical overlap. See figure 3.48.

Although it is possible, using stressor/susceptibility analysis, to determine possible problem sources in a circuit the failure probability of a component alone does not indicate possible problem sources. Therefore it will be necessary to trace influence factors having a dominant influence on a certain stressor/susceptibility overlap.

The main purpose of this project is to integrate reliability analysis in the design process of electronic circuits and systems. As mentioned earlier in this book this will require a (direct) relation between reliability and designable parameters. From the earlier paragraphs of this book it is possible to derive that, using stressor/susceptibility analysis, main influence factors for the reliability of a given component within a circuit are:

— Stressors

— Susceptibility

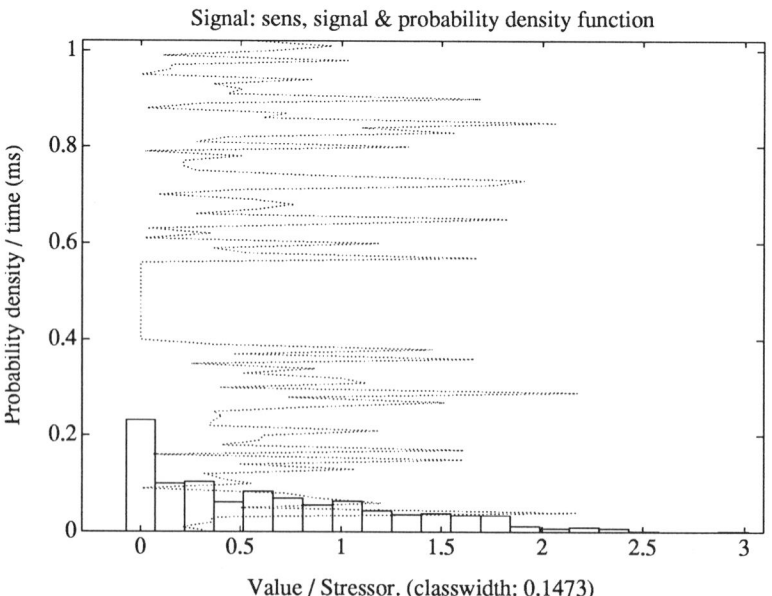

Figure 3.49: Direct relation stressors/signals

In many cases, as will be shown in Chapter 4, the susceptibility of the failure mechanisms in a component is a function of a number of component properties and a number of external factors. In those cases where the (circuit- and system-) designer has the freedom to select components on basis of these properties it is possible to use susceptibility aspects of a component as designable parameters. Chapter 5 will show the example of a Schottky diode where susceptibility aspects were used in this way. In most practical cases the design aspects of susceptibility related parameters will be the domain of the component designer.

An alternative to the use of susceptibility aspects in design optimization is the use of stressors for this purpose. Especially in those cases where stressor/susceptibility interaction is caused by certain peaks in the stressor function it will have considerable advantages to concentrate the optimization on the stressor function. This is also the case when alternative components with adequate susceptibility are not available.

The optimization of the reliability of a certain circuit concentrating on the stressor function is in many aspects more complicated than optimization using susceptibility aspects. When using the susceptibility aspect of a component in the optimization process the actual optimization will most times concentrate on the

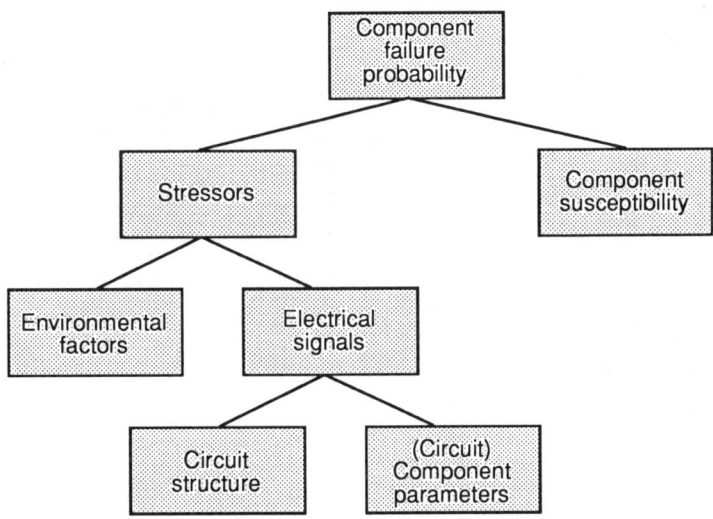

Figure 3.50: Relation failure probability/design para.

component itself. When using stressor functions in an optimization process it will be necessary to concentrate on all the external factors contributing to a certain stressor function. To simplify the optimization process the influence factors, contributing to a stressor function, are split into two categories

— Electric signals (and related parameters such as power dissipation, etc.)

— Other external influence factors such as ambient temperature, humidity and mechanical stresses.

Generally speaking a designer will have certain degrees of freedom to introduce changes in the first category. It is very important that influence factors in the second category are known in detail but, generally speaking, for most electronic circuits the possibility to introduce changes in this second category are limited.

The analysis step from stressor functions to electrical signals will require the reverse process that was used in the deriving of stressor sets. For the reverse analysis it is important to select those parts of electrical signals causing stressor/susceptibility overlap. See figure 3.49 for a simple, one-dimensional example.

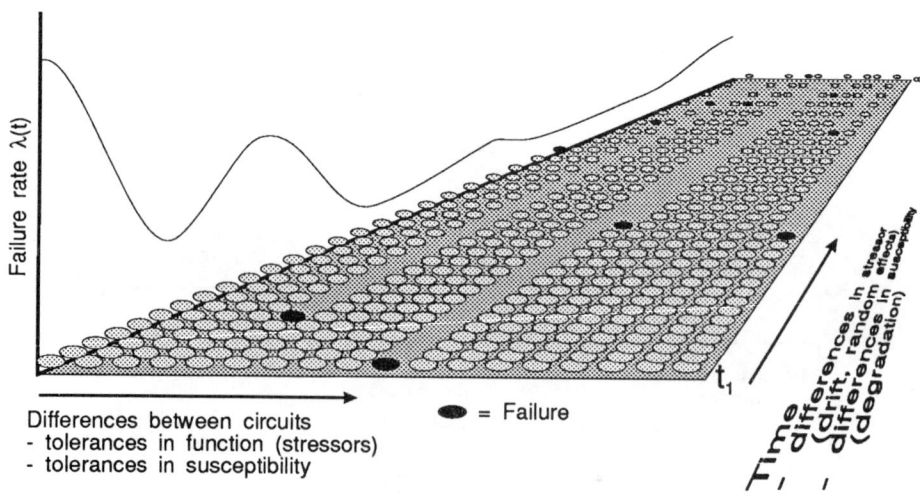

Figure 3.51: Time-dependent failure rates

To derive circuit parameters with a dominant influence on the selected electrical signal or electrical signal peak it is possible to use the well- known (tolerance) sensitivity analysis. Tolerance sensitivity analysis is used to describe the dominant effects of tolerances in part parameters on circuit level:

$$TS_x = \frac{\delta\ Y(x)}{\delta x} \qquad\qquad (3.60)$$

where Y(x) is a circuit signal (such as a node voltage) and x is a (part) parameter such as a resistor value.

Using this combination of stressor/susceptibility analysis and sensitivity analysis it is possible to give a direct relation between the reliability of components in a circuit and designable parameters. See figure 3.50. Disadvantage of the optimization method, presented in this section, is that the required calculations on the level of individual circuits will become, even for circuits of medium complexity, to complicated for practical optimization. Although this method demonstrates the existence of the wanted direct relation between reliability of a component and designable parameters on circuit level Chapter 6 of this book will show a method more suited for practical application.

3.7. **Summary of failure prediction**

Using stressor and susceptibility models it is possible to predict reliability of a component, based on the time-dependent distributions of both the susceptibility and the stressors. Using these models it is possible to relate effects of parameter drift, increasing susceptibility (degradation) and distributions of susceptibility and stressors to the well-known reliability and failure rate curves. See figure 3.51. Important difference between the presented method and the traditional "constant failure rate" curve is that the latter expressed in terms of the method discussed here assumes a population consisting of identical devices (no weak subpopulations) suffering under constant stressors. The well-known bathtub curve assumes, explained in terms discussed in this chapter, weak subpopulations failing during the first part of circuit lifetime and degradation-drift effects causing failures at the end of the circuit lifetime. Another important difference, also explained more in detail in Chapters 4 and 5 is that, speaking in stressor/susceptibility terms, main failure causes are often the peaks of the stressor function. Traditional reliability methods mainly use the mean stress on components. Chapter 5 will show that practical experiments indicate a majority of the failures being related to peaks in stressor distributions. Summarizing main differences between traditional reliability methods (see Chapter 2) and the presented reliability analysis method:

	Traditional method	Presented method
Component population	Identical components	Component distributions: functional, susceptibility
Stress	Average stress on identical components	Stressor function for individual components
Time-dependency	None: $\lambda = \lambda_c$ unknown: bathtub curve	Able to cover drift (functional) and degradation (susceptibility)

	Traditional method	Presented method
Influence factors	Mainly thermal and "environment/application class"	Detailed stressor/susceptibility interaction

Another aspect of the presented method is, as the last sections of this chapter have shown, the possibility to relate failure mechanisms directly to designable parameters. Although the use of this direct relation is possible, disadvantage is the very complex interaction structure. Therefore the presented relations are useful to prove a relation between reliability and designable parameters but hardly usable in practical situations. Chapter 6 will show a more simple method, derived from the presented stressor/susceptibility interaction, based on the principles of Monte Carlo analysis, which is usable in a simpler way in the design optimization process.

4
Deriving Susceptibility Modes from Failure Mechanisms

4.1. Introduction

One of the most important conclusions of Chapter 2 was the impossibility to use existing reliability prediction models in the process of circuit optimization. These models do not give the required relation between designable parameters and circuit reliability. Therefore reliability optimization will require other models. The main demand in this respect will be a relation between designable parameters and reliability.

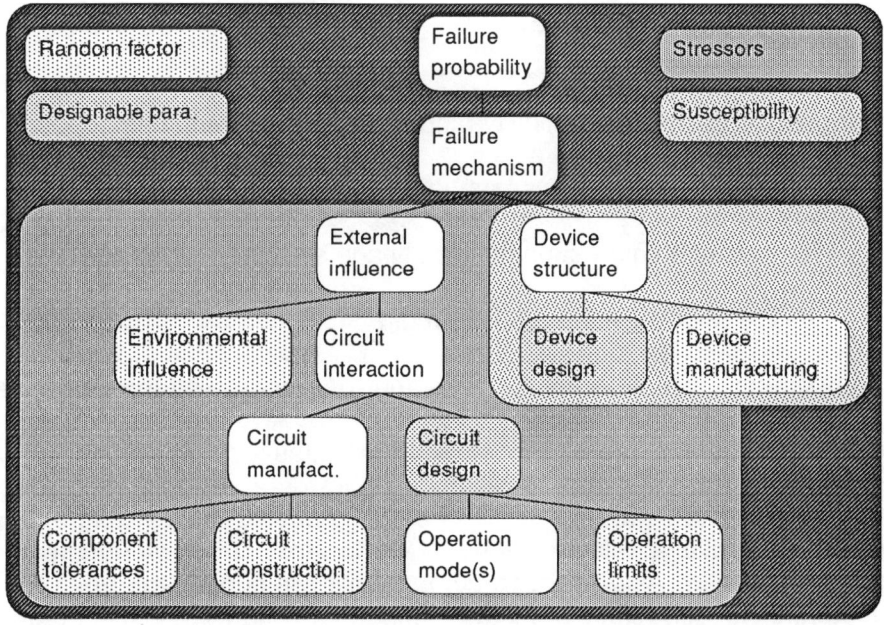

Figure 4.1: Influence factors on failure mechanisms

In Chapter 2 influence factors related to a certain failure mechanism were structured in accordance with figure 4.1. This chapter will concentrate especially on device-internal aspects of failure mechanisms; from the elementary failure mechanisms a set of susceptibility models will be derived. The stressors of such a susceptibility model will also be presented.

Chapter 3 gave a theoretical possibility for an analysis technique from which such a relation can be derived. In this chapter elementary possibilities for reliability optimization were also presented.

As a first stage in the reliability analysis and optimization process this chapter will derive susceptibility models (in accordance with Chapter 3) for electronic components. For that purpose the first part of this chapter will concentrate especially on fundamental failure mechanisms occurring in a wide range of electronic components. This overview of failure mechanisms is in no way complete. Many people have done research on a wide range of failure mechanisms occurring in a large variety of electronic components. The main purpose of the presented failure mechanisms is to present a method to derive stressor sets from physical failure mechanisms.

The second part of this chapter will derive, for three practical components, stressor/susceptibility models using these failure mechanisms. All these components have multiple failure mechanisms, each with multiple stressors. Appendix B will sketch outlines of susceptibility models for more straightforward components which have only one or two failure mechanisms with simple stressor sets.

This chapter will derive susceptibility models for the following components:

— Diode X (medium power schottky diode)

— Transistor Y (high-speed, high-voltage switching transistor)

— IC Z (motor drive integrated circuit)

The components were selected for a number of reasons. All the components are traditionally associated with reliability problems. Also the components have either widely differing reliability models (with often, as a consequence, different reliability figures) or the practical reliability figures are very different from the predicted failure rate figures. For all these components the existing reliability models do not give a relation between designable circuit parameters and component reliability. Also all these components are subject to rather complicated internal signals. Although all the components share the aspect that they are bipolar semiconductors, the presented method to relate physical failure mechanisms to stressor sets will also be usable for other classes of components.

4.2. **Failure mechanisms in electronic components**

Generally speaking there are two different categories of failure mechanisms applicable to electronic components. First there are failure mechanisms related to the electrical stress in a circuit. In this case electrical stresses are the direct cause of (often immediate) changes in the structure of a device, resulting in termination of the capabilities of a device for normal operation. In case of overstress the component is operated beyond its intended conditions.

Second there are failure mechanisms related to the intrinsic aspects of a component [Ame87]. Amaresekera indicates that due to flaws in the internal structure of a device it is possible that the internal properties of a device are changed in such a way that failures occur within the normal operation limits of a device. These failures may appear immediately after the production of the component. In that case it is often possible to associate these failures with imperfections in certain steps of the production process of devices. In other cases the failures are related to a combination of (history of-) operating conditions and manufacturing aspects. In those cases failures will not occur immediately but after a certain time, often quite dependent on the stresses applied during device operation.

The following sections will discuss the failure mechanisms existing in many electronic components. Important aspect of these sections is to give a relation between the failure mechanisms and stressors. In the second part of this chapter the discussed failure mechanisms are used to derive susceptibility models for the three sample components mentioned above.

4.3. **Electrical overstress failure mechanisms**

Many books on component reliability do not discuss electrical overstress (EOS) failure mechanisms in detail because often failures due to EOS are considered failures due to poor design. From a component specialist's view this might be right but from a circuit designer's point of view the situation is somewhat more complicated. The second half of this chapter, together with Chapter 5, will show that in practical applications a majority of the component failures is related to EOS. The following sections will discuss some fundamental EOS failure mechanisms, applicable to many components. Some more detailed EOS failure mechanisms will be discussed in the end of this chapter as a part of the discussion of failure mechanisms for sample components.

4.3.1. *Thermal considerations*

Temperature is for many physical processes, including most failure mechanisms in electronic components, a factor of great importance. Before going into details of the various failure mechanisms this section will give a (simplified) overview of thermal aspects of electronic components. For many electronic devices the device temperature is, often for a large part, determined by two important factors: the environmental temperature and the power dissipation in a device.

In most components where power is dissipated there is a thermal balance between the environmental temperature and the dissipated power. For those components where the heat is generated in a relative small volume, for example semiconductors where most heat is generated in the junction(s), it is possible to use quite simple thermal models.

Figure 4.2 shows the internal structure of a simple diode. The diode crystal is soldered on a substrate. On the anode side a bonding connects the crystal to the anode lead, on the cathode side of the crystal the substrate is directly connected to the cathode lead. The entire crystal is moulded in plastic. In this case it is possible to distinguish the thermal resistances presented in figure 4.3 [Tod70].

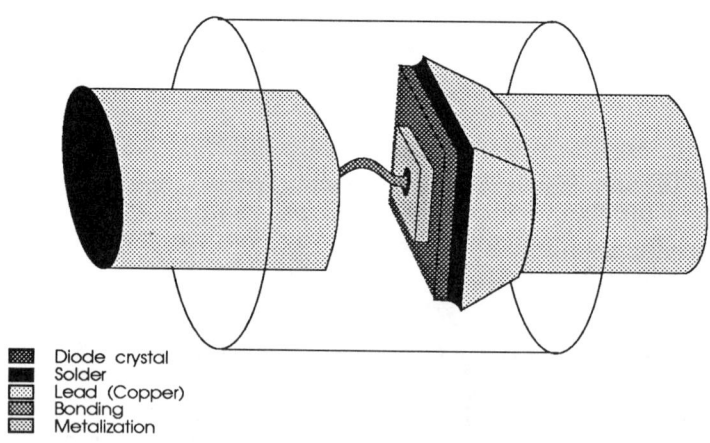

Diode crystal
Solder
Lead (Copper)
Bonding
Metalization

Figure 4.2: Internal structure of a simple diode

The device temperature for the static case can be derived from this model. In case of a given power dissipation and a given environmental temperature the junction temperature will rise until the heat flow through the total thermal resistance matches the heat flow caused by the internal power dissipation. The dynamic situation is somewhat more complex. Due to the thermal capacitances of the different materials figure 4.3 changes to figure 4.4. In those cases where the component is stressed using pulses, the temperature on a certain point of the component will depend strongly on the transient effects of the pulses, both in the electrical domain (power dissipation) and in the thermal domain (thermal resistances and capacitances). Only in those cases where the time-constant of

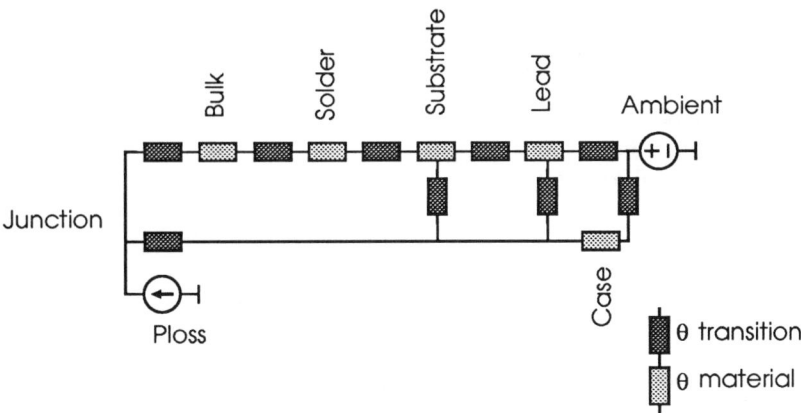

Figure 4.3: Static thermal structure of a simple diode

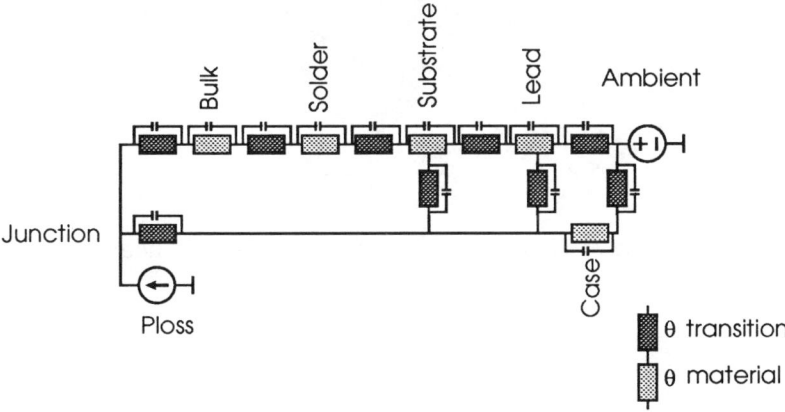

Figure 4.4: Dynamic thermal structure of a simple diode

the thermal network is in the same order of magnitude as the time-constant of the variations in power load will it be necessary to take time dependent thermal models into account.

For power devices it is often no longer possible to use models where the various parts of a device are modelled using one single thermal resistance. In those cases where temperature differences within materials are not relatively small it is necessary to use more accurate modelling techniques. See figures 4.5 and 4.6 for a practical example. The principle, however, remains the same: temperature distributions within a device are determined by the following factors:

— Dissipated power

— Environmental temperature

— Thermal resistances

— Thermal capacitances

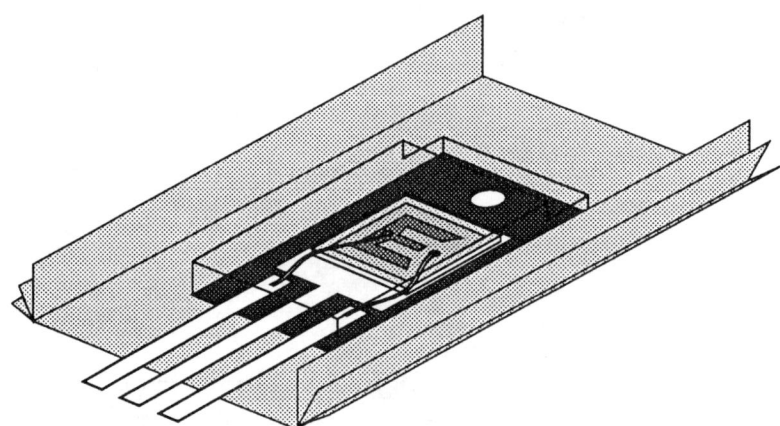

Figure 4.5: Simple transistor mounted on a heatsink

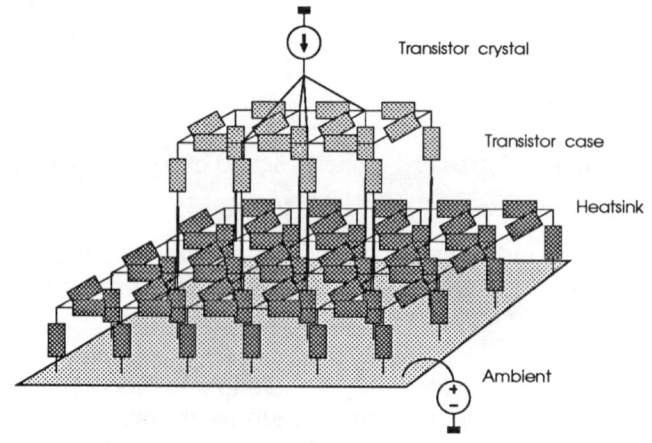

Figure 4.6: Simplified thermal structure

Figure 4.5 shows an example of a transistor mounted on a heat sink. Figure 4.6 shows a simplified approximation of the thermal resistances between the chip (the power source) and the thermal environment. Using these models it is possible to calculate the chip temperature as a function of the dissipated power, the environmental temperature and the various thermal resistances. A given maximum temperature in a device will lead to the following relation between allowed dissipated power and environmental temperature:

$$P_{max} = \frac{T_{dev_{max}} - T_{env}}{\theta_{device \to environment}} \qquad (4.1)$$

Figure 4.7 gives a practical example of a relation between environmental temperature, dissipated power and temperature at the source of power generation (semiconductors: junction).

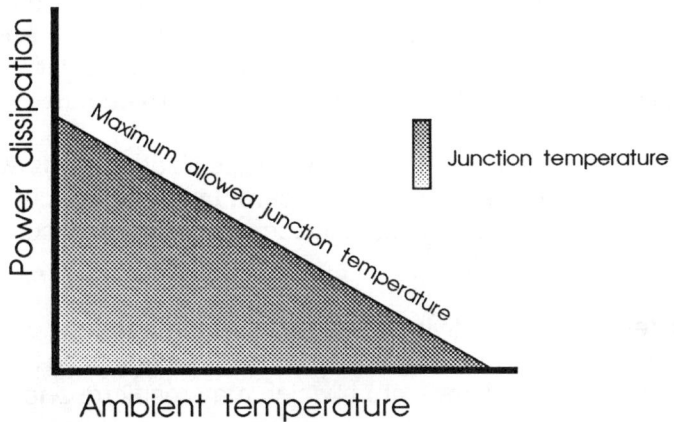

Figure 4.7: Relation between T_{amb}, P_{loss} and T_j

4.3.2. *Current breakdown (hot-spot melting)*

One of the best known failure mechanisms, valid for many components, is the effect of current breakdown. In all those cases where a current flows through a conductor with a certain electrical resistance this current will cause heat dissipation according to

$$P = I^2 R \qquad (4.2)$$

For a finite volume *dx dy dz* in a conductor the density of the local dissipated power will be:

$$\frac{dP}{dx\ dy\ dz} = J^2 \rho \tag{4.3}$$

The relation between both formulas is presented in the following expression

$$\tag{4.4}$$

$$P = \int_{x,y,z} J^2(x,y,z)\ \rho(x,y,z)dx\ dy\ dz$$

It is obvious that the power dissipation in a conductor is dependent on the current density, the volume of a conductor and the resistivity of the used material. From these expressions it is easy to derive that the power dissipation is maximal in those parts of the conductor where the current density is high or the resistivity is high. One of the causes of higher resistivity is, for example, the existence of dislocations in the material, which can cause electron scatter. Another reason for local increase of power dissipation is a decreased area of the conductor due to mechanical distortions of the material such as sharp angles.

The effect of increased power dissipation in localized regions will, generally speaking, be a rise in temperature. The temperature rise will depend, in practice, on the parameters of the thermal network such as the thermal conductivity from the (hot) spot to the ambient.

For many materials the resistivity has a positive temperature coefficient. This means that ρ rises for an increasing temperature. As a result the local dissipation will result in a higher local temperature which may result (due to the increased resistivity) in a higher local dissipation. In those cases where the heat flow to ambient is not able to stabilize this process (due to the increased local temperature the heat flow to ambient will also increase) the local temperature will rise until the conductor melts. Some examples of resistivity temperature coefficients and melting points are given in the table below [Wea87].

Material	Resistivity $(\rho(\Omega m\ 10^{-9}))$	$\frac{d\rho}{dT}$	Melting Point $(^\circ C)$
Aluminium	28.24	0.039	660
Copper	17.2	0.039	1083
Gold	24.4	0.034	1064
Silver	15.9	0.038	961

See the figures below for a practical illustration of current breakdown (More details are presented in Appendix B). It is possible to derive the following stressors for the mechanism of current breakdown:

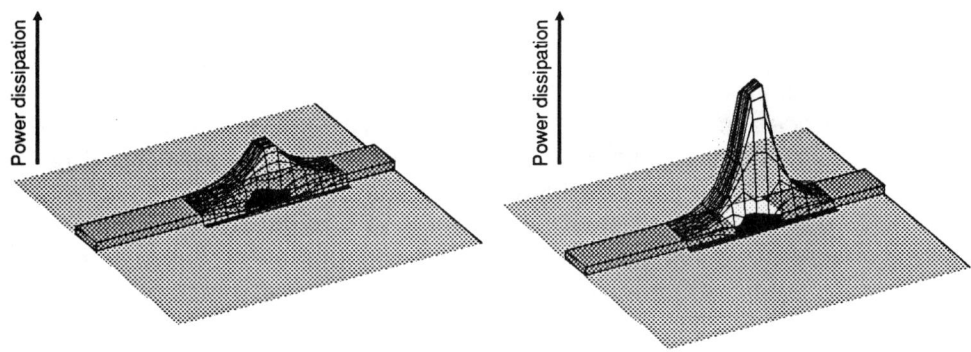

Figure 4.8: Distribution of power dissipation

Figure 4.9: Distribution power dissipation, positive dR/DT

Failure mechansism	Stressors
Current breakdown	Current density
	Environmental temperature

The following material aspects have influence on this failure mechanism:

— Resistivity of the material

— Impurities/mechanical distortions in the material causing increase in current density

— Thermal resistivity coefficient

4.3.3. *Power breakdown (thermal cracks)*

The temperatures of components are allowed to rise up to a level where physical processes change the properties of the component. For semiconductors this would mean that temperatures of several hundred degrees centigrade could be allowed before the actual structure within the silicon changes. At high temperatures the dopant atoms are able to move through the material thus changing the properties of the device. (See one of the following paragraphs: secondary diffusion.) Another theoretical limit is the intrinsic temperature of the silicon. Above a temperature of about 350-400 ºC silicon will behave as intrinsic material. As a consequence the junctions will loose their definition and the device is no longer operating [Mul86]. In practice, however, the thermal/mechanical structure of the device will be a very important factor in relation to the maximum temperature of a device.

An important problem in the thermal behaviour of an electronic component is related to the differences in thermal expansion of the different materials used. These differences in expansion coefficient cause mechanical stresses in a component which could ultimately lead to cracks in the component. Another problem is the finite thermal conductivity of the elements in the thermal circumference. Due to the various thermal resistances it is possible that within the material differences in temperature occur. In some cases this may also result in thermal cracks of the material. See figure 4.10 for a practical example.

Figure 4.10: Thermal cracks

This results in the following stressors for the failure mechanism of thermal breakdown (thermal cracks)

Failure mechansism	Stressors
Thermal breakdown (thermal cracks)	Dissipated power
	Environmental temperature

The following material aspects have influence on this failure mechanism:

— Thermal expansion coefficient of the materials

— Thermal resistivity coefficient of the materials

4.3.4. *High-voltage breakdown*

High voltage breakdown occurs, generally speaking, at the moment that due to a given (high) electric field, a current flows through an otherwise isolating layer of material. Due to the nature of the material and the nature of the electric field it is possible to distinguish several types of breakdown. According to Amerasekera it is possible to distinguish the following breakdown mechanisms for normally isolating materials [Ame86]

— Impact ionization

— Avalanche and Zener breakdown

— Electron-trap ionization

A common aspect of all voltage breakdown mechanisms is the presence of an electric field enabling charge carriers in the valence band to gain energy required for transfer to the conduction band. The difference between both mechanisms is related to the question how the actual conduction is initiated. The following paragraphs will discuss the different voltage breakdown failure mechanisms using the model of a simple capacitor. For semiconductors the dielectric material should be replaced by the (due to a reverse bias voltage) isolating depletion region.

4.3.4.1 *Impact ionization*

Impact ionization occurs in those cases where free electrons in the conduction band have gained sufficient energy to enable, in the case of collision, electrons in the valence band to be elevated to the conduction band. See figure 4.11 for

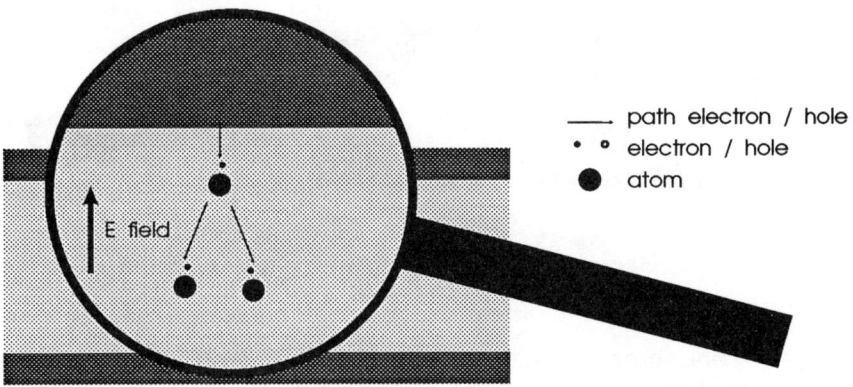

Figure 4.11: Impact ionization

an example. This failure mechanism depends mainly on the applied electric field as the electrons are accelerated by this field. Therefore the main stressor for this failure mechanism is

— Electric field

As impact ionization will cause disruptions in the structure of the material the effect of impact ionization will often be destructive.

4.3.4.2 *Avalanche and Zener breakdown*

In many respects avalanche breakdown is similar to the impact ionization described above. Important difference is that in many cases the impact ionization will cause irreversible crystal disruptions while in case of avalanche breakdown the process will not lead to immediate destruction. In case of avalanche breakdown a free carrier is able to gain enough energy to cause impact ionization. In this case the collision will result in two extra charge carriers: a hole and an electron. See figure 4.12. Temperature has minor influence on avalanche breakdown. Due to temperature rise the mean free path of electrons will decrease resulting in a decreasing probability that an electron will gain enough energy to cause ionization [Mul86a]. As a consequence the avalanche break-down voltage will have a slightly positive temperature coefficient [Klo89].

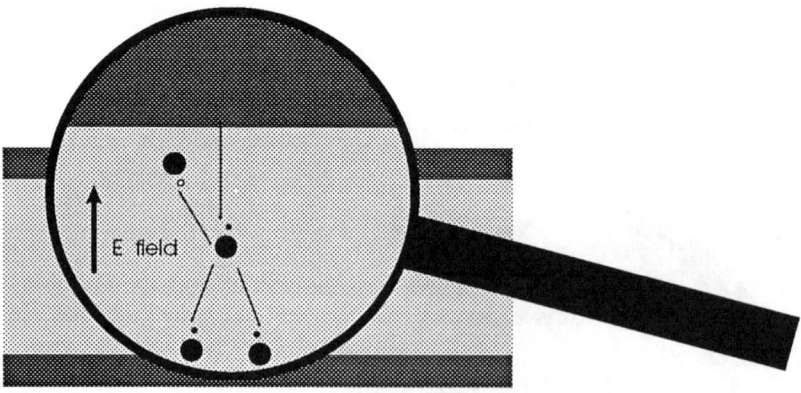

Figure 4.12: Avalanche breakdown

In the case of Zener breakdown the electric field itself is able to raise an electron from the valence band to the conduction band, thus creating two charge carriers: a hole and an electron. See figure 4.13. This breakdown effect is dependent on the applied electric field. The Zener breakdown voltage has a slightly negative thermal coefficient. As a consequence of rising temperature the probability of transition of an electron from the valence band to the conduction band will slightly increase under a given electric field. This will cause decreasing breakdown voltages [Mul86a].

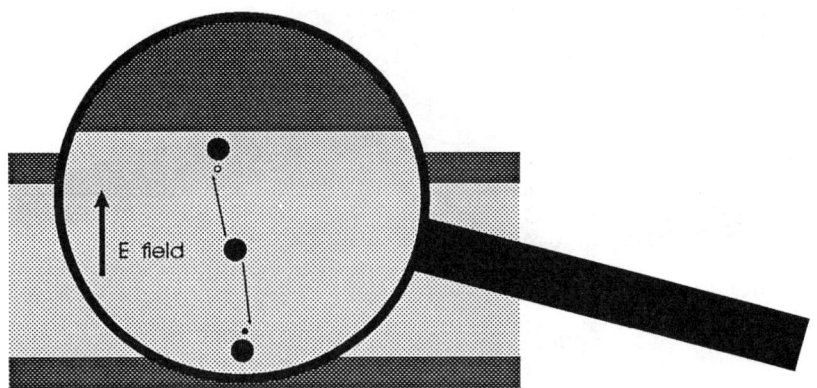

Figure 4.13: Zener breakdown

As a consequence it is possible to derive the following stressor for avalanche and Zener breakdown:

Failure mechansism	stressors
Avalanche and Zener breakdown	Electric field

Although these mechanisms are not necessarily destructive it often initiates a destruction process. Quite often breakdown will cause high currents combined with relative high voltages (= high dissipation) in devices. The resulting high temperature is in many cases sufficient to trigger one of the other failure mechanisms mentioned in this chapter as secondary failure.

4.3.4.3 *Electron-trap ionization*

In those materials where a certain disorder exists, due to for example impurities, another failure mechanism is possible [Bot85]. This impurity or hopping conduction occurs when particles "jump" from one impurity location to another. See figure 4.14. Amerasekera describes this failure mechanism as one of the two major breakdown mechanisms in MOS gate oxides [Ame86]. The difference between the breakdown mechanisms described above and this failure mechanism is that hopping breakdown is a relative slow phenomenon with a major temperature dependence. For this breakdown mechanism the following stressors are used:

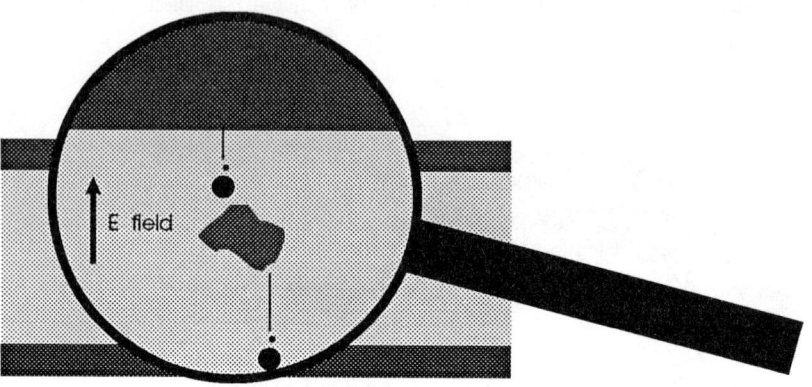

Figure 4.14: Electron trap ionization

Failure mechansism	stressors
Electron trap ionization	Electric field
	Temperature
	Time dependent behaviour of the electric field Static or quasi-static fields cause impact ionization

4.4. **Long term failure mechanisms**

At this moment a wide range of intrinsic failure mechanisms of electronic circuits is known. Especially for integrated circuits intensive research is performed to find the nature of failure mechanisms. This book will show some examples of failure mechanisms in electronic components where designable circuit parameters are involved. The following sections will discuss these failure mechanisms

— Corrosion

— Electromigration

— Secondary diffusion

The corresponding sections will not discuss the physical fundamentals of these failure mechanisms but will derive corresponding stressor sets from these failure mechanisms.

4.4.1. *Corrosion*

One of the problems for devices encapsulated in non-hermetic packages is the problem of corrosion [Ame87]. Due to corrosion metalization patterns in components may turn into an open circuit. See figure 4.15. Corrosion may occur due to a combination of:

— Moisture

— D.C. operating potentials

— Cl^- or Na^+ ions

Absence of one of these aspects will inhibit corrosion. As moisture is removed by increasing temperatures devices with a high power dissipation are less susceptible to corrosion. Stressors for this failure mechanism are

Failure mechansism	Stressors
Corrosion (long term mechanism)	Environmental temperature (negative influence on susceptibility)
	Dissipated power (negative influence on susceptibility)
	D.C. voltage

Although this failure mechanism is clearly time dependent the actual failure mechanism is so much dependent on random factors that it is hard to give a usable formula for the time dependency of this failure mechanism.

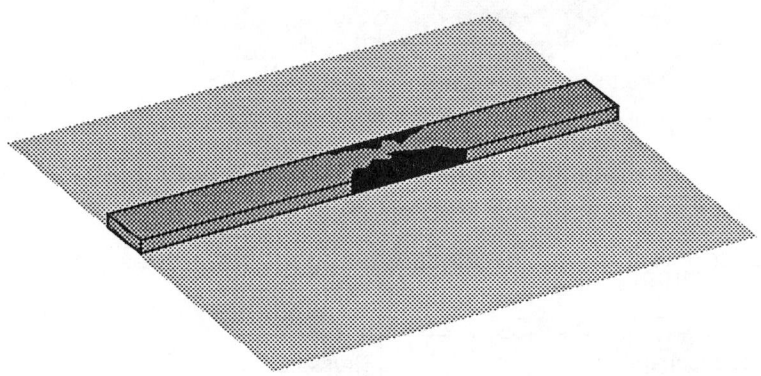

Figure 4.15: Corrosion in metalization patterns

4.4.2. *Electromigration*

In materials with sufficiently high current densities the continuous impact of electrons on the metal atoms may cause the atoms to move in the direction of the electron flow. Especially aluminium metalization tracks on semiconductor surfaces show this failure mechanism. This moving of metal particles may cause changes in the geometry of the conductor. Ultimately this may result in an open circuit. A void is created on one end of the track while the material piles up at the other side of the track. See figure 4.16. At this moment many people have studied this failure mechanism [Wad85] and a commonly accepted formula for the mean lifetime of metalization patterns is:

$$Mean\ lifetime = A\ J^{-n}\ e^{\Delta E / kT} \tag{4.5}$$

A constant depending on material and geometry

J current density

n constant

T average temperature of the conductors

k Boltzmann's constant

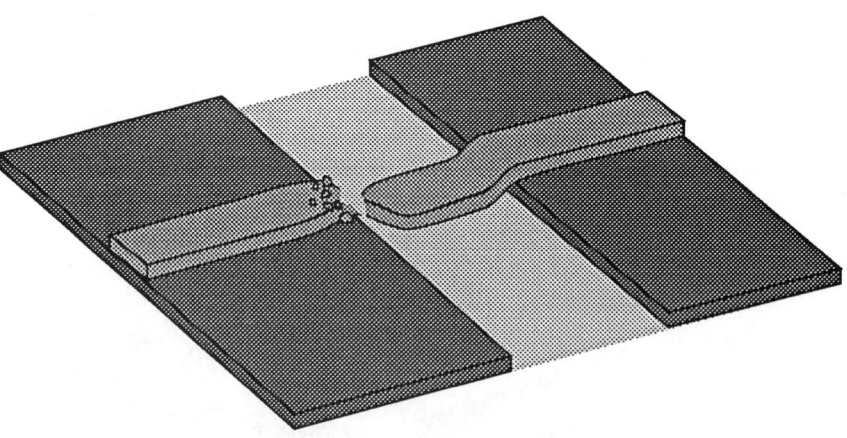

Figure 4.16: Example of electromigration

From this formula it is possible to derive that the main stressors for electromigration are:

Failure mechansism	Stressors
Electromigration (long term mechanism)	Current density
	Environmental temperature

4.4.3. *Secondary diffusion*

In many components the possibility exists of atoms of one element diffusing into another. This is especially the case for those materials where contacts between materials exist or materials where atoms of one material are already "built into" another material. A well-known case in this respect is secondary diffusion of semiconductors.

In the diffusion of semiconductors the doping profile depends on the concentration of dopant atoms during diffusion, the (high) temperature used during the diffusion process and the time during which the material is exposed to this high temperature, however, this does not imply that no diffusion processes take place at lower temperatures. Even in the absence of externally provided dopant atoms the internal dopant gradients are sufficient for reallocation of dopant atoms. See figure 4.17. The influence of this reallocation process is at room temperatures only relevant after very long times. Only in those cases where a crystal remains

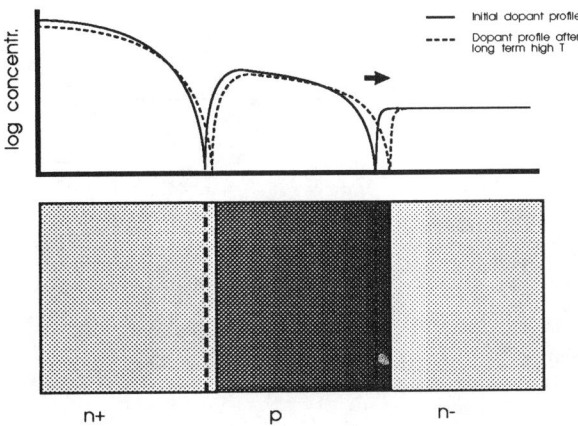

Figure 4.17: Secondary diffusion

for a long time at a high temperature (and no other failure mechanism causes problems earlier) effects of secondary diffusion can be expected. The exact behaviour in this respect is very dependent on the used dopants, the used concentrations and the thickness of the various layers. As this mechanism will cause, generally speaking, no immediate "hard" failures but a drift in parameters it is obvious that the term "failure mechanism" for this process is not quite satisfactory. The mentioned parameter drift might be a cause for other failures, inside or outside the component, and therefore it is mentioned in this chapter. Stressors for this failure mechanism are:

Failure mechansism	stressors
Secondary diffusion (long term mechanism)	Temperature

4.5. **Additional failure mechanisms for bipolar semiconductors**

The failure mechanisms mentioned in the previous sections are applicable to a wide range of electronic components. As the practical examples in the end of this paragraph deal with bipolar semiconductor devices this section will discuss some additional failure mechanisms existing in bipolar devices.

4.5.1. *Pulse power effects*

One of the previous sections discussed the failure mechanisms related to DC power dissipation in a device. For bipolar junctions there is an additional failure mechanism, also related to power dissipation [Wou86].

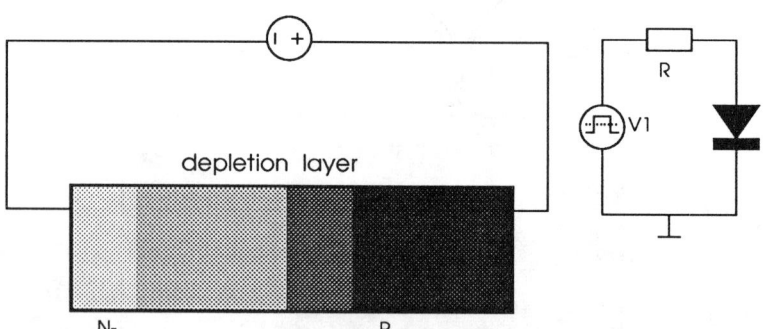

Figure 4.18: Reverse biased diode

Given a bipolar junction under reverse bias conditions. See figure 4.18. Changing the polarization of the bias voltage will result in a fast change of the polarization voltage for the diode. The diode will be able to reverse current immediately due to the fact that holes and electrons can start flowing from the diode terminals directly. Some time, however, is required before a steady state current is established. In the example given the time required is defined by the ratio of the charge difference between a steady state reverse biased diode and a steady state forward biased diode on one side and the applied current on the other side [Mul86b]. The voltage-time diagram, the current-time diagram and the power dissipation diagram are presented in the first part of figures 4.19 to 4.21 (switching from reverse bias to forward bias). The rapid formation of steady state forward charge conditions can be seen especially in figure 4.20.

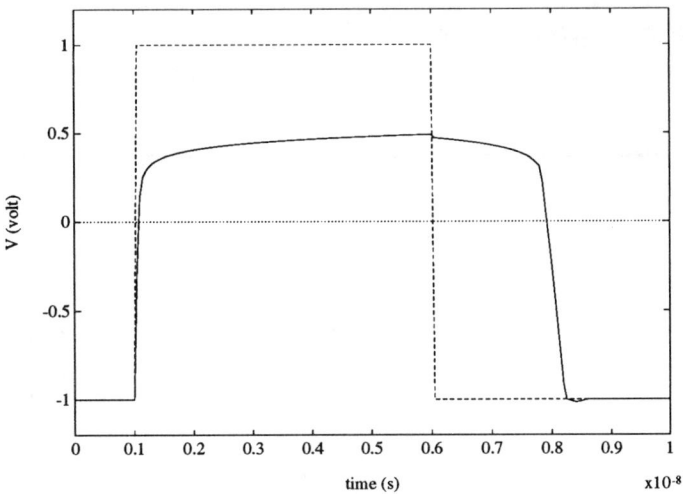

Figure 4.19: Diode voltage during pulse operation

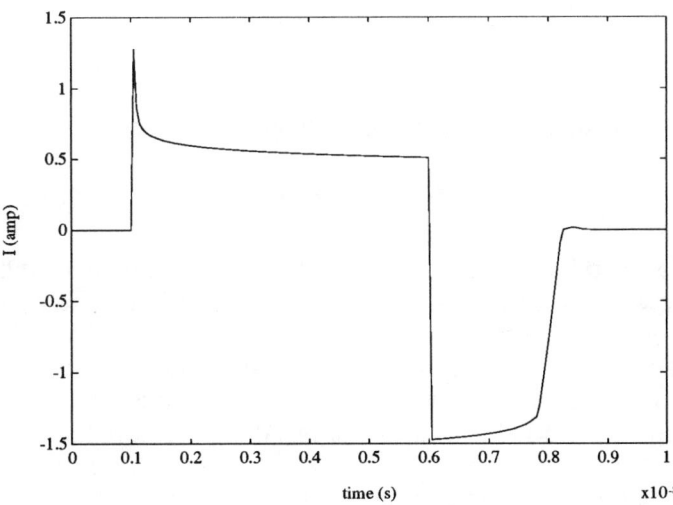

Figure 4.20: Diode current during pulse operation

Under extreme conditions this might either cause overcurrent problems or overpower problems. Stressors for this failure mechanism are:

Failure mechansism	stressors
Switch-on pulse power dissipation	Voltage slope $\dfrac{dV}{dt}$
	Current slope $\dfrac{dI}{dt}$

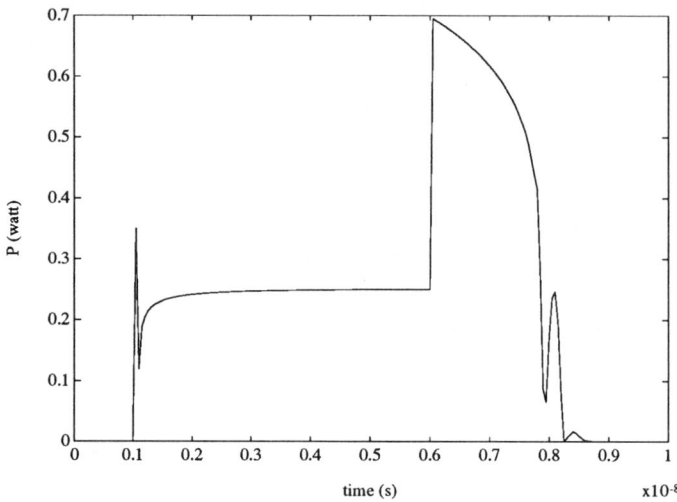

Figure 4.21: Power dissipation during pulse operation

Switching a forward biased diode to reverse has a somewhat different effect. Similar to the switch-on effect the diode is available to switch the current direction immediately, due to the availability of charge carriers at the terminals. The junction remains forward biased until the injected minority carriers near the edge of the depletion region are removed. This may result in considerable reverse currents through the diode for a certain time. At the moment the minority carriers are removed this current rapidly decreases. The voltage-time diagram, the current time diagram and the power dissipation diagram are presented in the second part of figures 4.19 to 4.21 (switching from forward bias to reverse). The maximum reverse current in figure 4.20 is largely determined by the circuit (see figure 4.18) according to

$$Id_{reverse,max} = \frac{V_1 + Vd_f}{R} \qquad (4.6)$$

where Vd_f equals the diode voltage before switching.

The number of minority carriers that has to be removed depends on the stored charge in the diode which, in its turn, depends on the applied forward diode current. This leads, for this failure mechanism, to the following set of stressors:

Failure mechansim	stressors
Switch-off pulse power dissipation	Stored charge Q_s in the diode at the moment of polarity reversal (related to the forward current in the diode at the moment of polarity reversal)
	Voltage slope $d\frac{V}{d}T$
	Maximum reverse diode current
	Applied reverse voltage

All these terms contribute to the power dissipated during reverse conduction. As can be seen in figure 4.21 the reverse dissipated power may considerably exceed the power dissipated under forward conduction. An important aspect of this failure mechanism is the fact that this effect is closely related to the thickness of the crystal. As the crystal becomes thicker more charge is stored in the device and therefore not only the turn-on time and the turn-off time of the diode increase but also the dissipated pulse power. Therefore high-voltage diodes will be more susceptible to this failure mechanisms than low-voltage diodes. This aspect contrasts with the often used derating policy: "the higher the (voltage) rating the better".

4.5.2. *Second breakdown*

Transistor operation is limited by a number of parameters. As for all devices, these are the maximum current, voltage and power dissipation ratings. The power dissipation is a function of the operation mode of the transistor. From the previous sections it is possible to derive a DC failure model for the bipolar junction transistor using three failure mechanisms:

— Voltage breakdown

— Power breakdown

— Current breakdown

These three failure mechanisms are combined in a so-called DC Safe Operating Area (SOA) diagram. See figure 4.22. The shaded area in figure 4.22 indicates the allowed combinations of voltage and current in a transistor. The section on thermal aspects of breakdown mechanisms (4.3.1) showed that in case of pulsed operation, under the condition that the time constants of the thermal network are large compared to the time constants of the applied power pulses, it is allowed

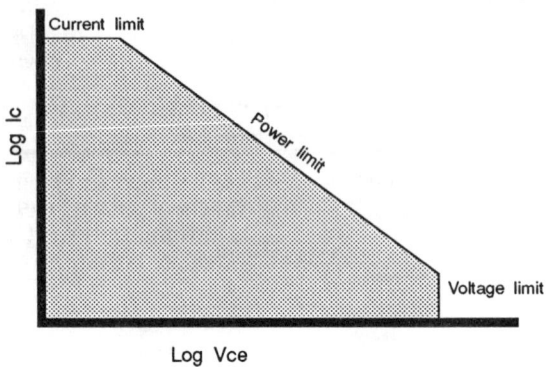

Figure 4.22: Three failure mechanisms for BJTs

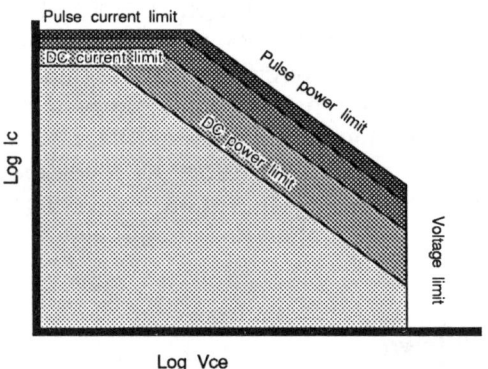

Figure 4.23: Enhanced failure mechanism model for BJTs

to use the peak power in calculations of crystal temperatures (see also Appendix B). This implies that in the case of a given maximum junction temperature it is allowed to use pulsed power peaks beyond the normal DC allowed power maximum. This changes figure 4.22 into figure 4.23. As both the power break-down and current breakdown failure mechanisms are thermally dependent the graph should be extended into a third dimension incorporating thermal SOA aspects. See figure 4.24. The use of such charts explains historically that designers have taken (often erroneous) the "biggest" transistor to be safe.

In high power and high frequency applications within maximum current and voltages ratings catastrophic failures occurred. These failures increased when employing the derating strategy. Apparently, another transistor mode exists:

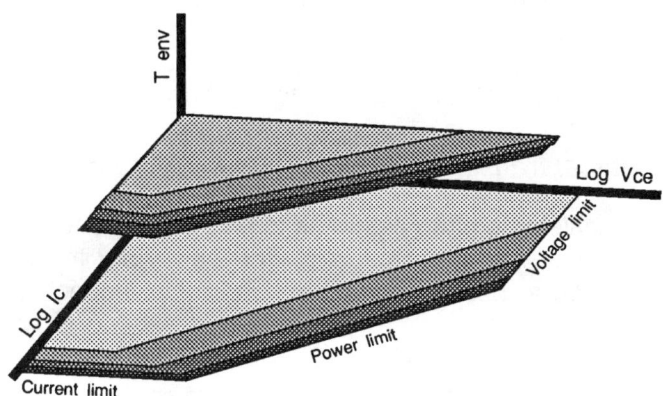

Figure 4.24: Three failure mechanisms including temperature

These failures manifested themselves by a sudden collapse of the collector emitter voltage and loss of control of the base drive. Microscopic analysis reveals local melting of both silicon and contacting materials. Obviously, the crystal was subjected over a very small area to considerable heating caused by extreme concentrations of collector current. Historically this effect was called: "Second breakdown" (SBD) [RCA70] [Let84]. Second breakdown is found in various

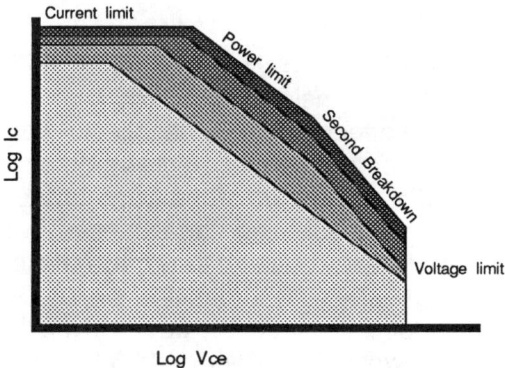

Figure 4.25: Failure mechanism model including SBD

technologies and under both forward-bias and reverse-bias conditions. However, the failure mechanisms for both bias conditions are different.

The following sections will give a short overview of second breakdown failure mechanisms and will derive the relevant stressor sets. A detailed discussion of the physical aspects of second breakdown is given by Humphreys [Hum88].

These sections will concentrate on the failure mechanisms for second breakdown in bipolar power transistors. As second breakdown effects are closely related to geometrical aspects of transistors first a brief explanation of the effect of the geometrical transistor structure on the switching behaviour will be given.

4.5.2.1 *Geometrical transistor aspects related to breakdown effects*

Theoretically a transistor is often assumed to be a homogeneous device having one emitter, one base and one collector. The behaviour in all parts of these terminals is assumed to be identical. Problem in this respect is especially the

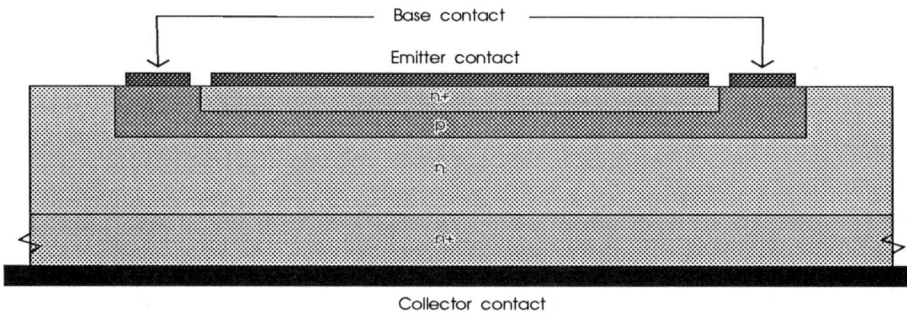

Figure 4.26: Cross section BJT

construction of the base of a transistor. See figure 4.26 for a cross-section diagram of a simple n^+pnn^+ transistor. Due to the ohmic effects of the base channel combined with the effects of the basecollector capacity, the base of the transistor will not behave homogeneously. This inhomogeneous behaviour depends on the distance of a given part of the base to the base electrode. See figure 4.27. Although the resistors and capacitances in this figure are by no means ideal components, it is obvious that they may have considerable influence on the transistor switching, especially for fast transients. During switching on the current will concentrate especially under the edges of the transistors emitter. This effect is called "current crowding". See figure 4.28.

During switching off the transistor will remain conductive longer in the middle of the emitter area than at the edges. The charge, stored in the transistor base, is first removed at the edges and later in the middle of the base channel. See figure 4.29. As a consequence the current through the transistor will "pinch-in" in the middle of the emitter area. This effect is quite similar to the high reverse current in diodes immediately after re-polarization. As long as (a part of-) the junction is forward charged it remains current carrying.

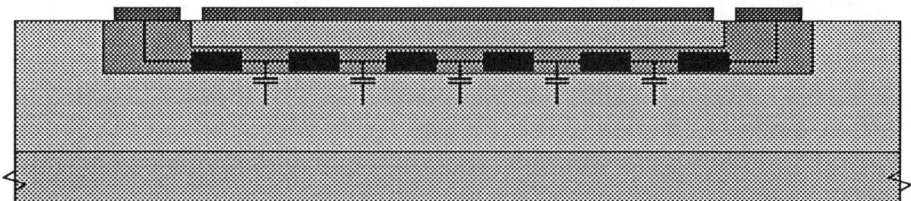

Figure 4.27: Non-homogeneous switching of the base

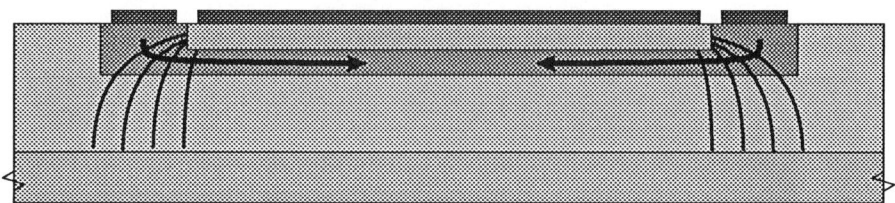

Figure 4.28: Current crowding during switch-on

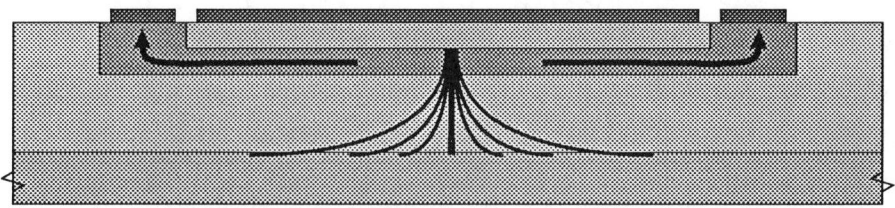

Figure 4.29: Pinch-in during switch-off

How these current crowding and pinch-in effects influence transistor behaviour can be demonstrated using a square planar transistor. See figure 4.30. It is possible to simulate the behaviour of a large, inhomogeneous transistor using an array of small homogeneous transistors and model the effects of non-homogeneous switching using a base network. Although the accuracy of such a model is insufficient for detailed simulation it is useful as an indication of areas where current crowding or pinch-in will occur. See figure 4.31.

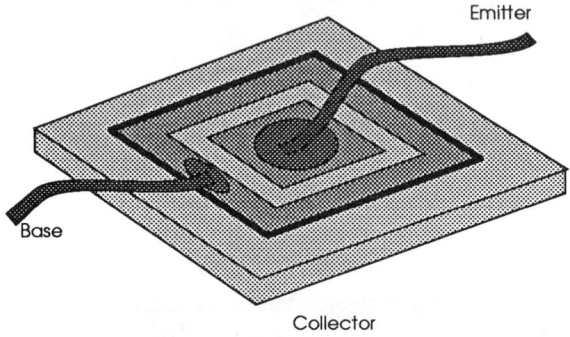

Figure 4.30: Square transistor used to simulate SBD

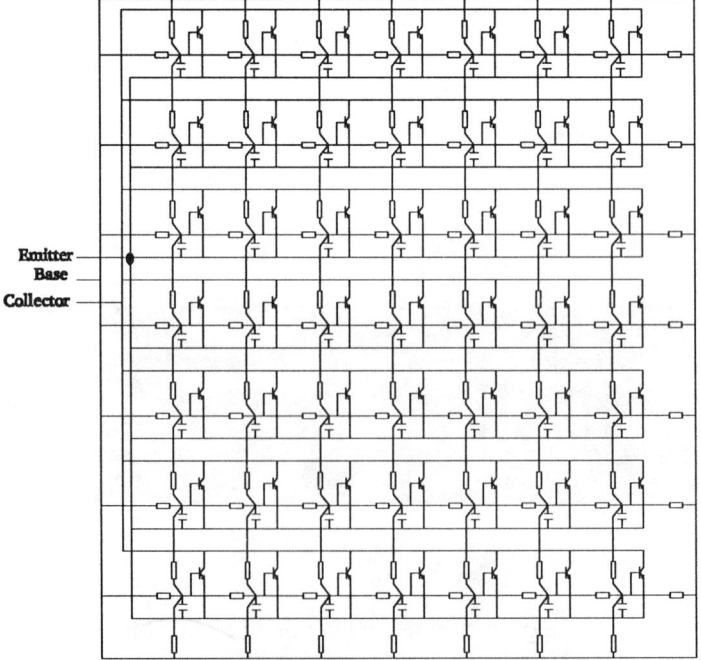

Figure 4.31: Model of finite element transistor

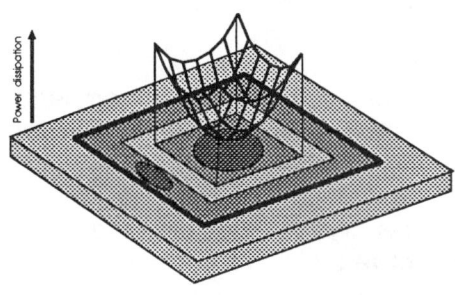

Figure 4.32: Geometrical dependency of current crowding

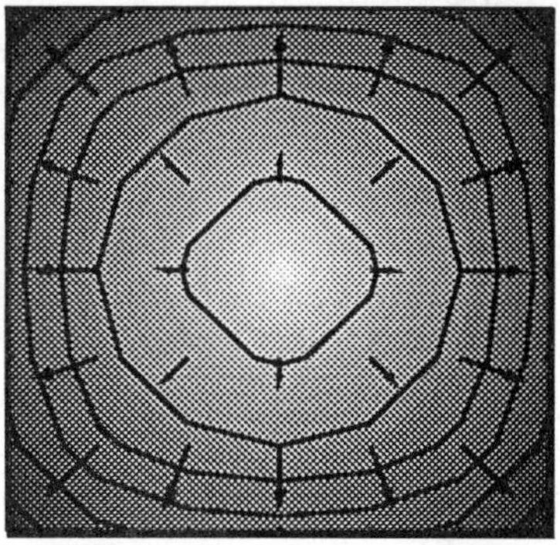

![Figure 4.33](Distribution of power dissipation figure)

Figure 4.33: Distribution of power dissipation

Switching-on this transistor will result in the power distribution given in figures 4.32 and 4.33. From these figures it is possible to derive that especially the extremities of the emitter area of the transistor are susceptible to current crowding effects.

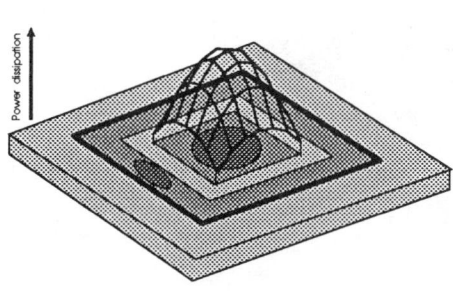

Figure 4.34: Geometrical dependency of pinch-in

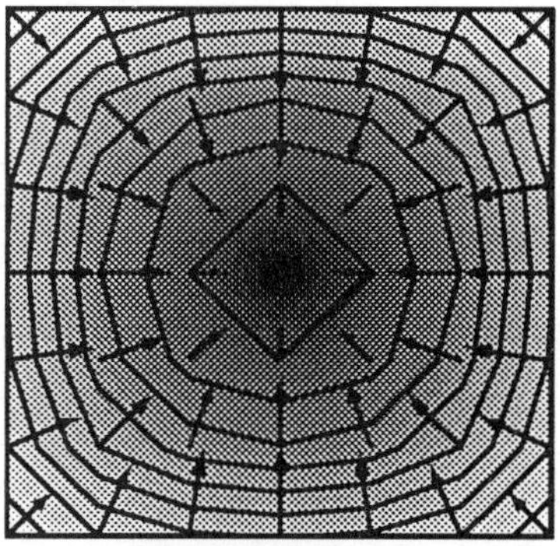

Figure 4.35: Distribution of power dissipation

Switching off the transistor will result in the power distribution given in figures 4.34 and 4.35. From these figures it is possible to derive that especially the interior of the emitter area of the transistor is susceptible to pinch-in effects.

4.5.2.2 *Forward-bias second breakdown*

Because of the forward-bias condition the so- called current "crowding" effect manifests. See figure 4.28. This effect concentrates the collector current all around the emitter periphery, therefore power dissipation will concentrate in the extrimities of the transistor.

Due to the localized heating and the negative thermal coefficient of V_{be} the base-emitter voltage decreases and injection increases in the depletion region (temperature coefficient of 2 mV/ °C).

With the local increase in current density caused by the localized temperature rising and enhanced thermal generation of carriers a positive feedback loop emerges which can become unstable at some critical temperature. In this case a "hot-spot" will be formed. A hot spot can be caused by a defect of some sort such as a thin spot in the base layer or as a crystal defect. As the hot spot becomes hotter and more electrically conductive additional current flows to the hot spot increasing power regeneratively, resulting in the destruction of the device.

A typical example of forward second breakdown is given in the figure 4.36. In many cases practical failures will not show such a clear picture. As one of the secondary effects of failure is a short-circuit between collector and emitter the failure effects of this secondary failure may be a completely burned crystal.

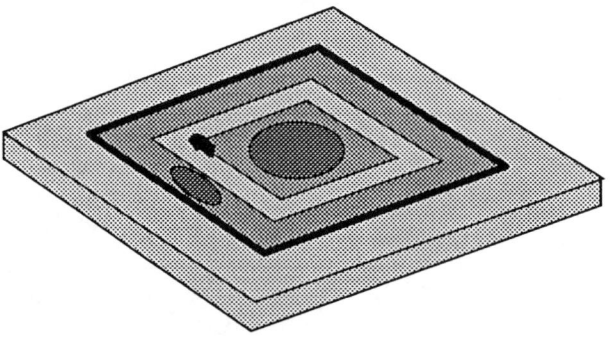

Figure 4.36: Burn-mark caused by forward bias SBD

This gives the following stressors for forward-bias second breakdown.

Failure mechansism	Stressors
Forward second breakdown	Collector emitter voltage
Forward second breakdown	Slope of the base current during switching-on $d\frac{I_b}{dt}$
	Slope of the collector current during switching-on $d\frac{I_c}{dt}$
	Environmental temperature

The following device aspects are related to this failure mechanism

— Conductivity of the base channel

— Geometry of the transistor

— Doping profiles, especially of the transistor base, the collector and the transistor substrate

4.5.2.3 *Reverse-bias second breakdown*

Under reverse-bias conditions the injected collector current is concentrated in the central portions of the emitter. This so-called "pinch-in" effect is caused by the lateral base current through which the voltage across the emitter-base junction drops and constricts the current to the centre of the emitter. See figure 4.29 for an example using a simple two-dimensional transistor.

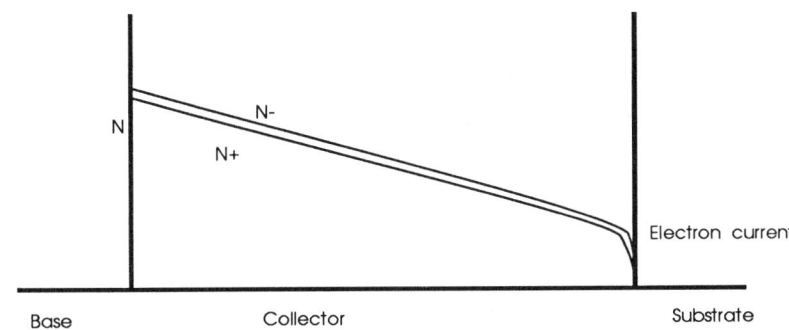

Figure 4.37: Distribution of collector charge carriers

Figure 4.37 shows the distribution of electrons and holes in a cross section of the collector of a conducting transistor. The base region is extended into the collector area. Directly after polarity reversal a current will remain flowing (see also section 4.5.1 on diode switching). In the same time an Efield will build up at the collector- substrate junction. See figure 4.38.

Under some conditions (high initial collector current, high collectoremitter volt-age) this Efield reaches a level where avalanche breakdown occurs. See figure 4.39. In this case the Efield collapses and breakdown occurs.

The previous mentioned "pinch-in" effect undermines the derating strategy and increases the reverse bias second breakdown effect. Especially when switching inductive loads the current will remain at a constant level directly after switching the transistors base. Combination of a constant current and a transistor already

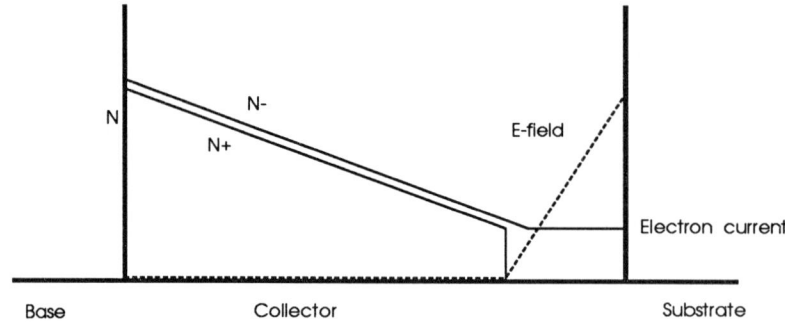

Figure 4.38: Rising *E*field after base-switching

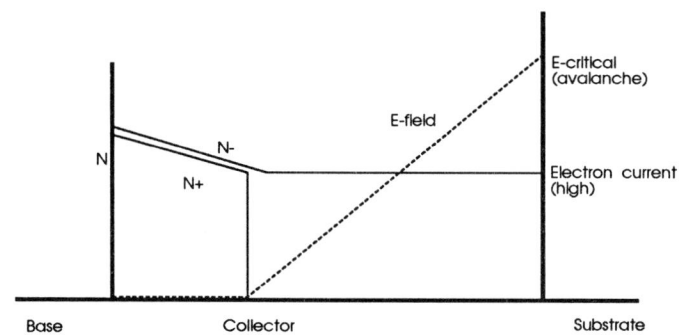

Figure 4.39: Critical *E*field (V_{ce}, I_c)

partially switched off will result in a considerable increase in current density. A "bigger" (larger emitter area) transistor suffers from larger voltage drops across the emitter-base junction and hence from more concentration of the current to the center of the emitter. [*]

One of the solutions to prevent pinch-in effects seems a rapid discharge of the transistor base. There is, however, an important limitation in this respect. A very rapid discharge of the transistor base will cause a remaining "charge bubble" under the middle of the transistors emitter area. Rapid discharging may cause a complete charge removal at the edges of the transistors base channel. In those areas where charge is completely removed the lateral conduction of the base

---

[*] For this reason many power transistors have finger structures. The use of finger structures prevents large single emitter areas.

channel drops, thus leaving a remaining charge under the middle of the transistors emitter area. See figure 4.40. Therefore it is important that the base discharge rate remains close to an optimum.

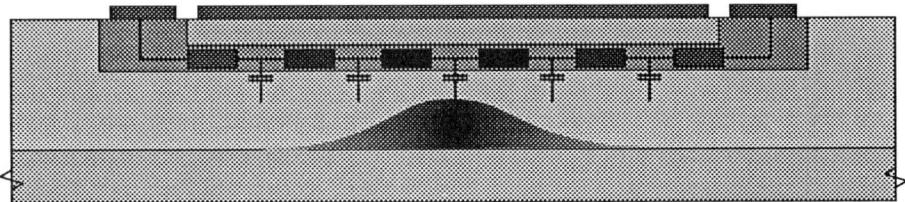

Figure 4.40: Removing base-charge too fast

A typical example of reverse second breakdown is given in the figure 4.41. In many cases practical failures will show a more damaged transistor crystal. As one of the secondary effects of failure is a short-circuit between collector and emitter the failure effects of this secondary failure may be a completely burned crystal.

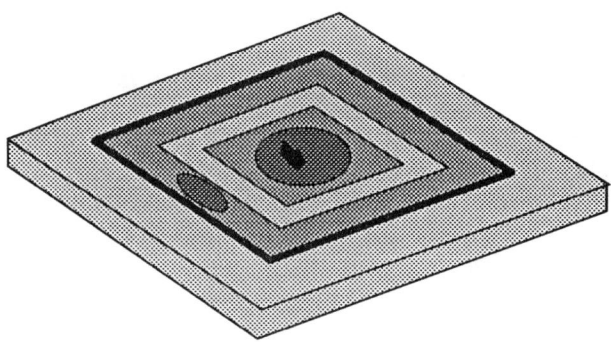

Figure 4.41: Burn-mark caused by reverse bias SBD

Based upon the bias conditions, one can make assumptions about the relative damage of the device under second breakdown conditions. Under forward-bias conditions the crowding effect allows a hot spot to carry only a small part of the total current. Under reverse-bias conditions the "pinch-in" effect makes a hot spot carry the whole current. Therefore, the damage of a device under reverse-bias condition will be more serious than of a device under forward-bias condition.

Together this gives the following stressors for reverse-bias second breakdown.

Failure mechansism	stressors
Reverse second breakdown	Collector-emitter voltage
	Stored charge at the moment of transistor switch-off (closely related to collector current at the moment of switch-off)
	Discharge speed $d\frac{lb}{dt}$ (optimum value)
	Environmental temperature

The following device parameters are related to this failure mechanism:

— Conductivity of the base channel

— Geometry of the transistor

— Doping profiles, especially of the transistor base, the collector and the transistor substrate

4.5.3. *Summary of the discussed failure mechanisms*

The previous sections have discussed a number of failure mechanisms occurring in electronic components. Although the presented failure mechanisms by no means cover the total spectrum of failure mechanisms in electronic components it is (intended as) a framework for the development of susceptibility models of practical components. The following table gives an overview of the discussed failure mechanisms with the related sets of stressors.

Failure mechanism	Stressor
Current breakdown	Current density
	Environmental temperature
Thermal cracks	Dissipated power
	Environmental temperature

Failure mechanism		Stressor
High-voltage breakdown,		
	(Impact ionization)	Electric field
	(Avalanche and Zener)	Electric field
	(Electron trap ionization)	Electric field
		Environmental temperature
		Time dependent behaviour of the electric field Static or quasi-static fields cause impact ionization
Corrosion (time dependent)		Environmental temperature (negative influence on susceptibility)
		Dissipated power (negative influence on susceptibility)
		D.C. voltage
		Moisture
Electromigration time dependent		Current density
		Environmental temperature
Secondary diffusion (time dependent)		Power dissipation
		Environmental temperature
Switch-on effect (diodes)		Re-polarization speed $\dfrac{dV}{dt}$
		Speed of charging $\dfrac{dI}{dt}$
Switch-off effect (diodes)		Stored charge Qs in the diode at the moment of polarity reversal
		Re-polarization speed $\dfrac{dV}{dt}$
		Maximum reverse diode current
		Applied reverse voltage
Forward bias second breakdown (bipolar transistors)		Collector emitter voltage
		Slope of the base current during switching-on $\dfrac{dIb}{dt}$
		Slope of the collector current during switching-on $\dfrac{dIc}{dt}$
		Environmental temperature

Failure mechanism	Stressor
Reverse bias second breakdown (bipolar transistors)	Collector-emitter voltage
	Stored charge in the transistor collector current at the moment of transistor switch-off
	Discharge speed $\frac{\mathrm{d}I}{\mathrm{d}t}$ positive influence: *fast discharging of the transistor base* negative influence: *too fast discharging of the transistor base will cause "charge bubbles".*
	Environmental temperature

Common aspect in many failure mechanisms is the thermal influence. The only cases where temperature is not a dominant influence factor are some of the high-voltage breakdown mechanisms. An important common aspect in the switch-off behaviour of bipolar semiconductors is the charge stored in the device and the (time-dependent) way in which this charge is removed.

4.6. Susceptibility models for practical components

The following sections will apply the failure mechanisms discussed above to some practical components. The main purpose is the development of a set of stressor/susceptibility models usable in practical situations.

4.6.1. *Diode X (Schottky diode)*

First component discussed in this section is the diode X (Schottky diode). Figure 4.42 gives a structural sketch of this component. This diode is a component intended for rectifying purposes. From the previous sections it is possible to derive a set of failure mechanisms, applicable to this component:

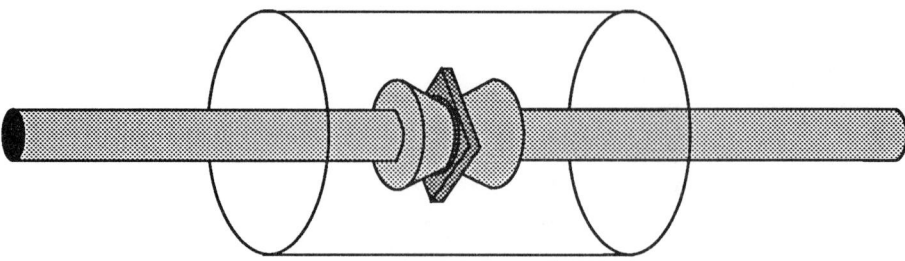

Figure 4.42: Structure diode X

— Current breakdown

— Power breakdown

— Avalanche breakdown

— (Pulse power effects)

4.6.1.1 *Pulse power effects*

The last term in the list, pulse power effects, is deliberately placed between brackets as the pulse power effects on Schottky diodes are somewhat different from the pulse power effects on standard np junction diodes. Due to the fact that the active diode area consists of the silicon-metal contact the discharge behaviour will be fast compared to discharge of np diodes. See figures 4.43 to 4.45 for the charge- and discharge- behaviour of this Schottky diode. From this graph it is possible to derive that, to prevent possible harmful effects of pulses under both charge- and discharge- conditions, a voltage rise of 10^9 V/s and a current rise of $0.5 \cdot 10^9$ A/s should not be exceeded.

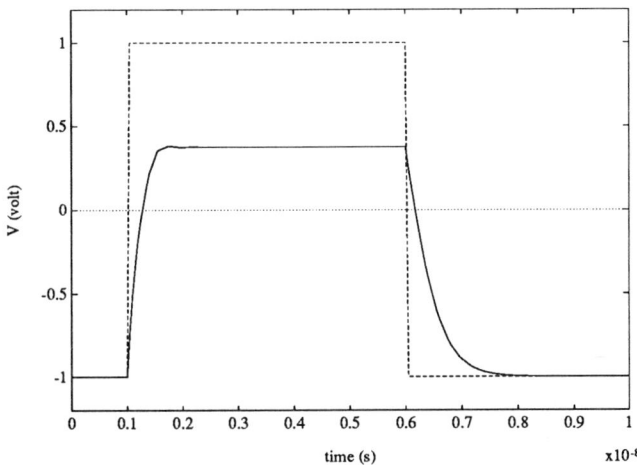

Figure 4.43: Diode X; voltage rapid switch-on/off

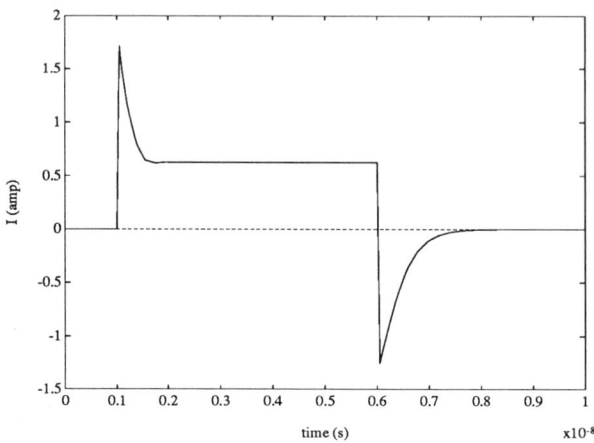

Figure 4.44: Diode X; Current rapid switch-on/off

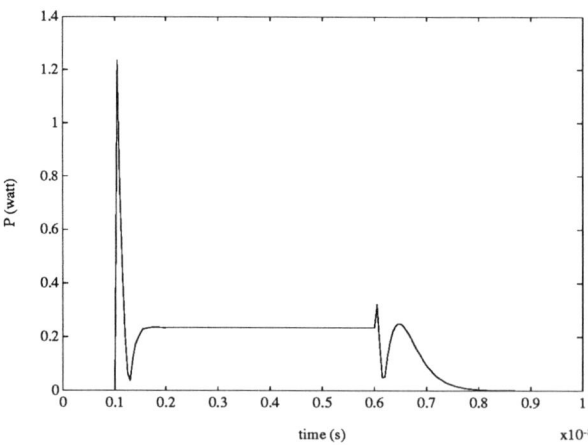

Figure 4.45: Diode X; power loss during rapid swiching

4.6.1.2 *Current breakdown*

For the current breakdown effect it is necessary to check the internal structure of the diode for any possibilities for current limiting effects in bondwires and/or whiskers. As can be seen in figure 4.42 the diode is directly connected to the leads. A consequence of this structure is that it is hardly possible to determine situations where current breakdown is a dominant factor. This is especially due to the good (thermal) conductivity of the relative thick copper leads from crystal to ambient.

4.6.1.3 *Avalanche breakdown*

The voltage breakdown behaviour for this device is guaranteed to be, according to the manufacturer, over 30 V. Own observations showed a distribution of voltage breakdown for this component as presented in figure 4.46. [*] A majority of the population fulfils the manufacturer's specification of 30 V allowed reverse voltage, however, an interesting sub- population exists with a breakdown voltage of about 5V. Although the reverse voltage breakdown of diodes does not necessarily mean diode destruction, operation of the diode in this area should be avoided.

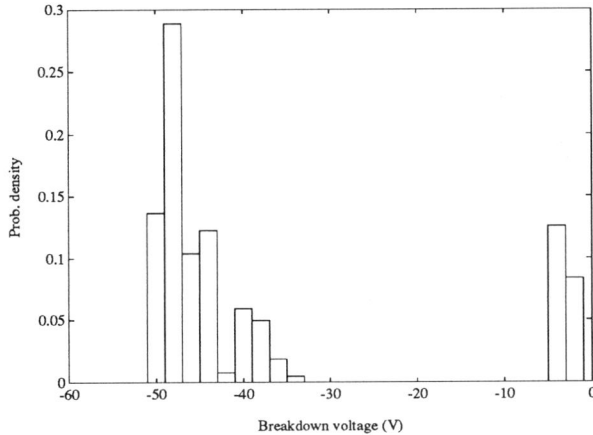

Figure 4.46: Distribution of breakdown voltage diode X

4.6.1.4 *Power breakdown*

Research on the power aspects of this diode , performed by Philips CE CIRG [Sch86] , shows that a maximum junction temperature ($T_{j,max}$) of 115 ^oC is allowed. This is derived from the internal structure of the diode, demands for thermal stability of the diode, the thermal resistance from ambient to junction ($\theta = 67$ K/W long leads, 72 K/W short leads) and a maximum ambient temperature of 60 ^oC. This temperature can be used to calculate the maximum forward and reverse combinations of current and voltage.

Summarizing the susceptibility limits for this diode are:

---

[*] Measurements on breakdown behaviour of Schottky diodes were performed on basis of several random sample series for a total of 2500 diodes by Mr Gommers of Philips CE in Eindhoven as part of this project.

Parameter	Susceptibility limit
V reverse	V breakdown (see distribution of breakdown behaviour)
T crystal	115 ^{o}C
$\dfrac{dV}{dt}$	$10^9 V\!/s$
$\dfrac{dI}{dt}$	$0.5 \; 10^9 A\!/s$

4.6.2. *High voltage transistor Y*

Second component discussed in this section is the transistor Y, a high voltage switching transistor. Figure 4.47 gives a structural sketch of this component. From the previous paragraphs it is possible to derive a set of failure mechanisms, applicable to this component:

— Current breakdown

— Power breakdown

— Avalanche breakdown

— Forward bias second breakdown

— Reverse bias second breakdown

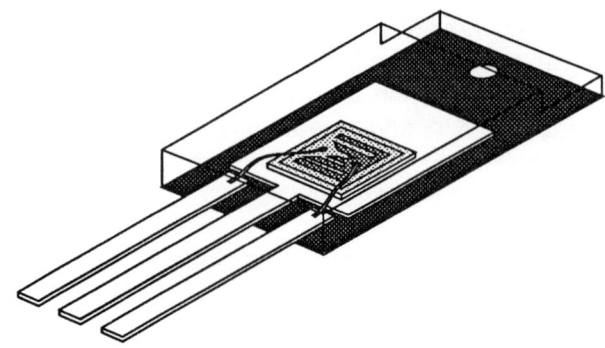

Figure 4.47: Structure of transistor Y

4.6.2.1 *Current breakdown*

From the internal structure of the device it is possible to derive that current breakdown is possible at two locations: the base bondwire and the emitter bondwire. For both wires aluminium bondwire is used. Tests at Philips in Hazelgrove showed that current breakdown of these bondwires does not occur below a peak current of 10 A. Important problem in determining exact values for this failure mechanism is the difficulty to avoid other failure mechanisms. At the mentioned current it is quite difficult (due to saturation effects in the transistor) to limit power dissipation thus avoiding other failure mechanisms.

4.6.2.2 *Power breakdown and secondary diffusion*

As mentioned earlier in this chapter there are two effects in which the temperature by itself might have an important influence on the failure probability of the device. First there is the probability of thermal cracks, second there is the long-term effect of secondary diffusion. Tests at Philips Hazelgrove showed that both failure mechanisms do not have a relevant influence below a junction temperature of 150 ºC.

4.6.2.3 *Avalanche breakdown*

In contrast with diodes transistors have multiple ways of showing avalanche breakdown. Important is the state of the transistor base. The most simple form of avalanche breakdown is avalanche breakdown with emitter-base short circuit. See figure 4.48. In this case the base-collector diode behaves like a standard diode under reverse bias conditions.

In case of an open base the transistor behaves like two diodes in series, the base-emitter diode under forward bias conditions and the base-collector diode under reverse bias conditions. See figure 4.49. As both diodes share the same anode, the charge in the forward diode will increase the susceptibility for breakdown in the reverse biased diode.

The breakdown voltage under open base conditions is specified as > 450 V. Practical measurements[*] show the distribution given in figure 4.50. The breakdown voltage under base-emitter short circuit was in all cases over 1000 V.[**]

---

[*] Practical measurements were performed on 200 transistors at the measurement department of Philips Hazelgrove and at Twente University as part of this project.

[**] Due to limitations in measurement equipment measurements over 1000 V were not possible.

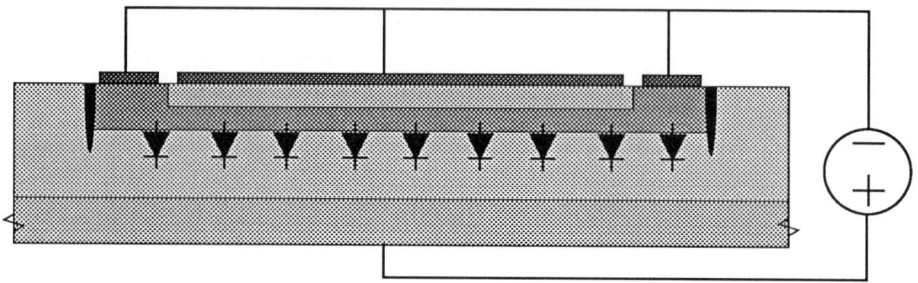

Figure 4.48: Breakdown voltage base short-circuit B$_{vces}$

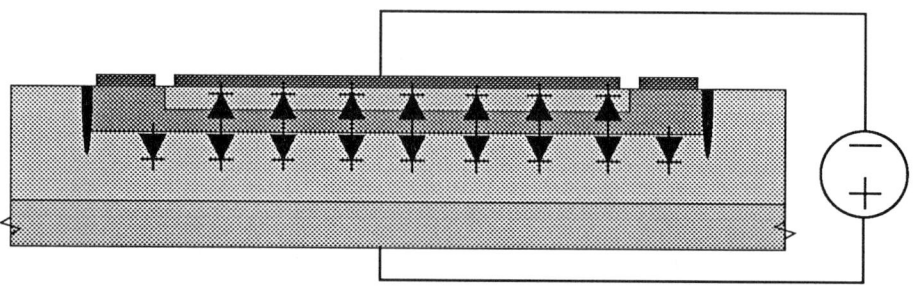

Figure 4.49: Breakdown voltage open base B$_{vceo}$

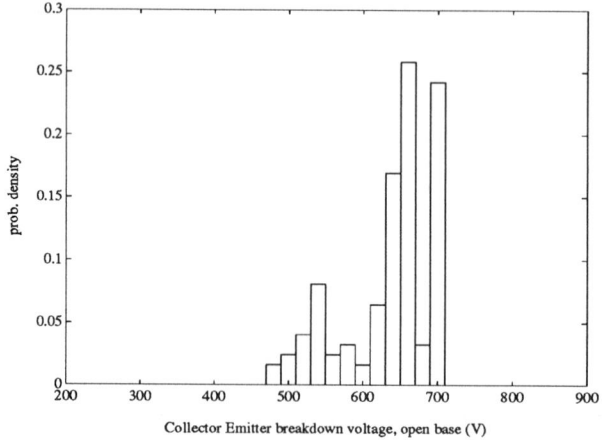

Figure 4.50: Distribution B$_{vceo}$ transistor Y

4.6.2.4 *Forward bias second breakdown*

Main functional demand to prevent forward bias second breakdown is to switch the transistor on fast. Important in this respect is the demand that the transistor should conduct homogeneously. The maximum switch-on time using a base current of 0.5 A is specified for this transistor as 1 s. To prevent additional delays from the driving circuit it is possible to define the minimum base current slope during switching-on as $0.5 \ 10^{6A}/s$. As mentioned in the previous sections, forward second breakdown depends also on the collector-emitter behaviour of the transistor. In those cases where the transistor switches a pure inductive load this inductive load will prevent initially excessive collector currents thus eliminating the possibility of forward second breakdown.

4.6.2.5 *Reverse bias second breakdown*

As mentioned in the previous paragraphs reverse bias second breakdown prevention requires an optimum switch-off time depending on collector-emitter voltage (after switching-off), collector current (before switching off) and the environmental temperature. Experiments at Philips Hazelgrove have resulted in the following relation between parameters [Wou86a]:

$$\frac{d\,Ib}{d\,t} = K(V_{ceo})\ Ic \qquad (4.7)$$

With

$K = 5\ 10^5 s^{-1}$ for V_{ceo} 400 V,
$K = 1.5\ 10^5 s^{-1}$ for V_{ceo} 700 V

For a V_{ceo} of 400V and a collector current of 900mA this would lead to an optimum slope of $4.5\ 10^{5A}/s$. Own observations have learned that failures occur in those cases where K is in the order of magnitude of $3\ 10^4 s^{-1}$. [*] Due to the inaccuracy of the observations a susceptibility limit of $K = 5\ 10^4 s^{-1}$ is assumed for a V_{ceo} of 400V. Observations at Philips Hazelgrove have shown that switching off too fast occurs at K values of about $1.5\ 10^6 s^{-1}$ for a V_{ceo} of 400V. Own experiments were not able to destroy transistors using rapid switching off. Summarizing the susceptibility limits for this transistor are:

Parameter	Susceptibility limit
Current breakdown	10 A
Crystal temperature	150 ^{o}C

---

[*]   Accurate observations with the available measurement equipment were not possible. Only possibility was to observe where failures did not occur.

Parameter	Susceptibility limit
Avalanche breakdown	450 V (open base) 1000 V (base-emitter short circuit)
$\dfrac{d\,Ib}{dt}$ (switch on)	$0.5\ 10^6 A/s$ (slower causes failures)
$\dfrac{d\,Ic}{dt}$ (switch on)	Not applicable in case of inductive load (most applications use inductive load)
$\dfrac{d\,Ib/dt}{Ic}$ (switch off)	$K = 5\ 10^4 s^{-1}$ (slower causes failures) $1.5\ 10^6 s^{-1}$ (faster causes failures)

4.6.3. *Integrated circuit Z (motor driver IC)*

One of the major differences between the components discussed in the sections above and this component is the situation that this component is an integrated circuit. Consequently it will not be possible to separate all the desired voltages and currents in order to determine a complete stressor/susceptibility model. Nevertheless this section will try to derive a stressor/susceptibility model which covers an important part of this component.

As discussed in Chapter 2 integrated circuit Z is a component intended for motor-drive functions in a video cassette recorder. See figure 4.52 for a blockdiagram of this circuit.

It is important to distinguish three different parts in this circuit: a digital part, an analog part and the power stages. The majority of the interactions within the two first parts take place within the circuit. The behaviour of the third section is, for a large part, also determined by the outside world. This is also the part of the circuit where a majority of the power is dissipated. Therefore especially the power stages are interesting for the development of stressor/susceptibility models. In the rest of the circuit the temperature is, for a large part, determined by the power dissipated in the power stages. As the internal interaction in the integrated circuit is hardly accessible for external measurements and is hardly influenced (on the level of failure mechanisms) by external parameters other than the environmental temperature the major part of this section will concentrate on the three power stages in this circuit.

In the previous sections susceptibility models were discussed for bipolar diodes and transistors. The table below gives an overview of the components used in the power amplifiers, the failure mechanisms relevant for these components and the way the stressors, relevant for these failure mechanisms, are related to external signals of the integrated circuit. See also figure 4.51.

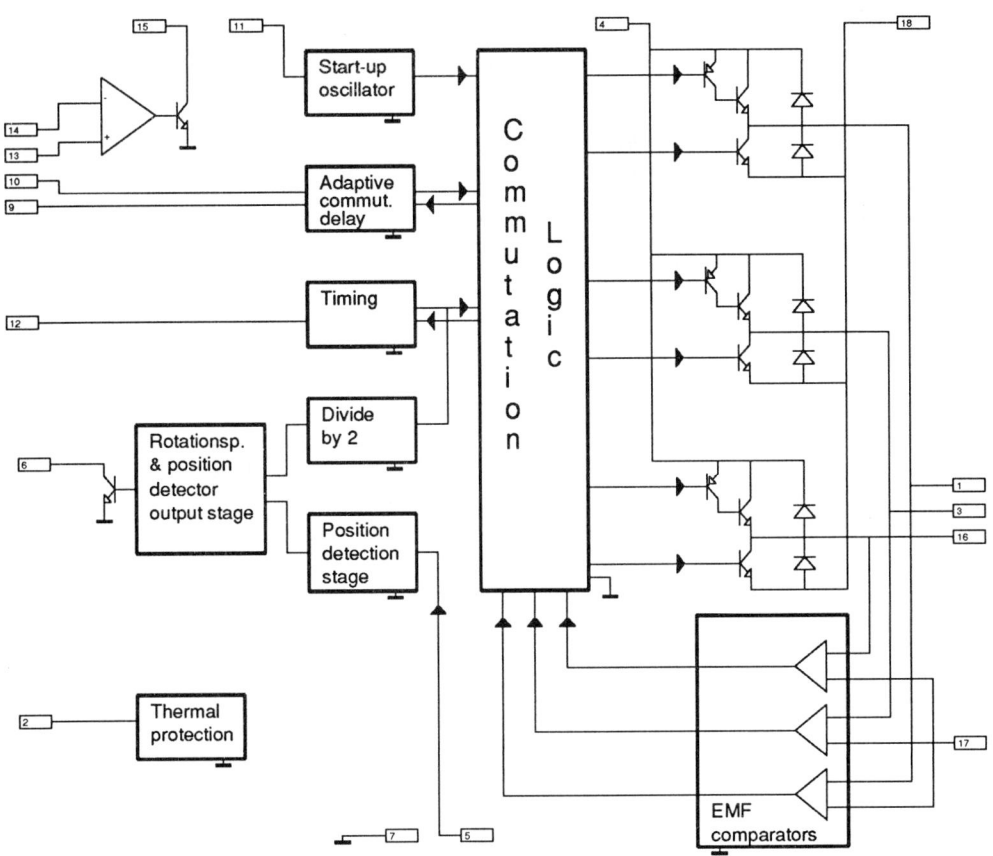

Figure 4.51: Internal structure of IC Z

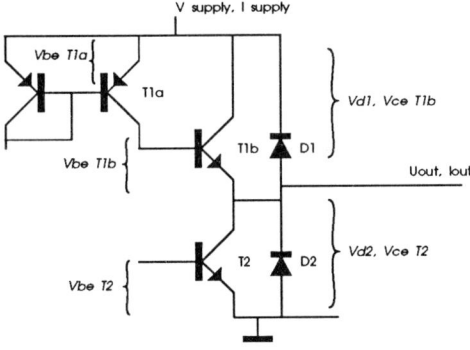

Figure 4.52: One of the power amplifier stages of IC Z

Component	failure mechanism	Stressors	Related circuit signals
Entire chip	current breakdown and electromigration (bondings, metalization patterns)	current density	Terminal currents*, Internal currents*
	power breakdown and secondary diffusion	junction (chip) temperature	power dissipation (V_{supply}, I_{supply})
Diode D1	Avalanche breakdown	V_{d1}	V_{supply}, V_{out}
	Switch-on power loss	$d\dfrac{V_{d1}}{dt}$, $d\dfrac{I_{d1}}{dt}$	V_{supply}, V_{out}, I_{out}
	Switch-off power loss	Q_s ($\sim I_{d\,0-}$), $d\dfrac{V_{d1}}{dt}$, $\dfrac{d\,I_{d1}}{dt}$, V_{0+}, I_{0+}	V_{supply}, V_{out}, I_{out}
Diode D2	Avalanche breakdown	V_{d2}	V_{out}
	Switch-on power loss	$\dfrac{d\,V_{d2}}{dt}$, $\dfrac{d\,I_{d2}}{dt}$	V_{out}, I_{out}
	Switch-off power loss	Q_{stored} ($\sim I_{d\,0-}$), $\dfrac{d\,V_{d2}}{dt}$, $\dfrac{d\,I_{d2}}{dt}$, V_{0+}, I_{0+}	V_{out}, I_{out}
Transistor T1b	Forward second breakdown	$V_{ce,T1b}$, $\dfrac{d\,I_b}{dt}$	V_{supply}, V_{out}^*, I_{out}, $I_{base,T1b}$
	Reverse second breakdown	Q_{stored} ($\sim I_{c\,0-}$), V_{ce}, $d\dfrac{I_b}{dt}$	V_{supply}, V_{out}^*, I_{out}, $I_{base,T1b}$
Transistor T2	Forward second breakdown	$V_{ce,T2}$, $\dfrac{d\,I_b}{dt}$	V_{out}, I_{out}, $I_{base,T2}^*$

Component	failure mechanism	Stressors	Related circuit signals
Transistor T2 (cont.)	Reverse second breakdown	$Qstored\ (\sim Ic_{0-})$, V_{ce}, $d\dfrac{I_b}{dt}$	V_{supply}, V_{out}, I_{out}, $I_{base,T2}$

Ibase T1b[*], X$_{0-}$[**], X$_{0+}$[***]

The first problem is: how to derive the parameters having influence on failure mechanisms from circuit signals. Therefore it will be necessary to go into some functional details.

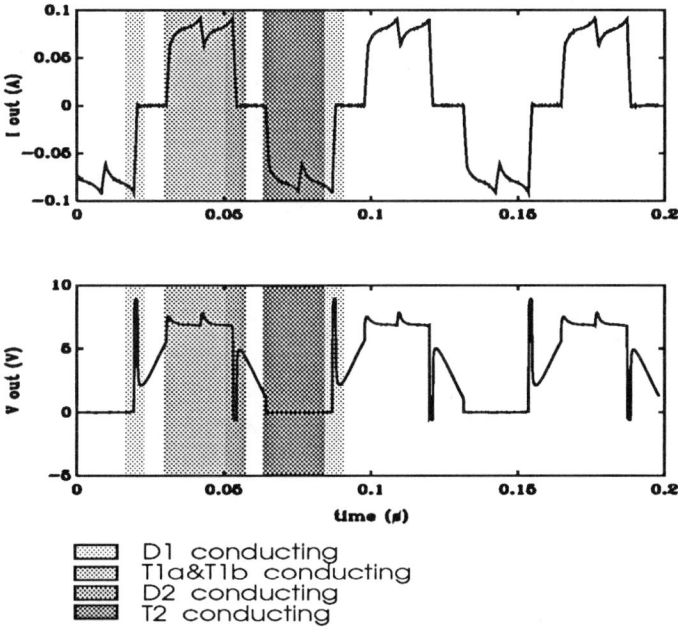

D1 conducting
T1a&T1b conducting
D2 conducting
T2 conducting

Figure 4.53: Output currents and voltages power stage

---

[*]   Internal IC signals

[**] X_{0-} denotes the state of parameter X at the time directly before polarity reversal

[***] X_{0+} denotes the state of parameter X at the time directly after polarity reversal

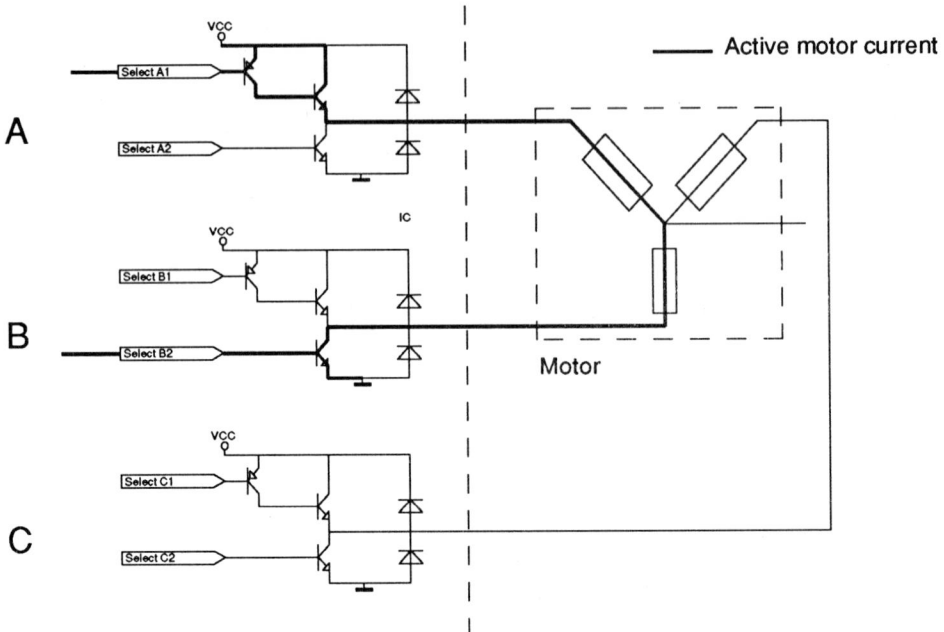

Figure 4.54: Currents through power stage IC Z

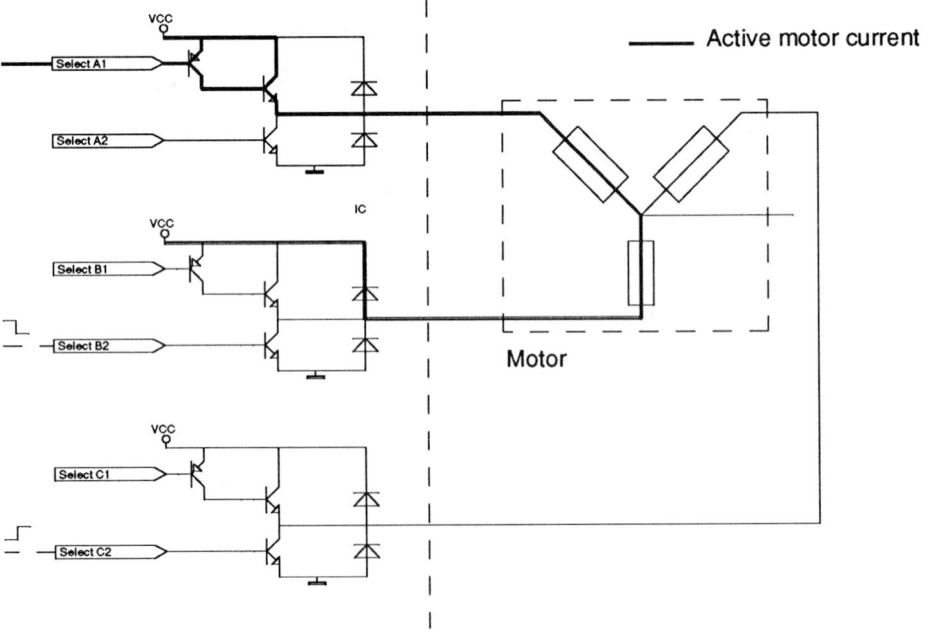

Figure 4.55: Use of flybackdiode D2 during switching

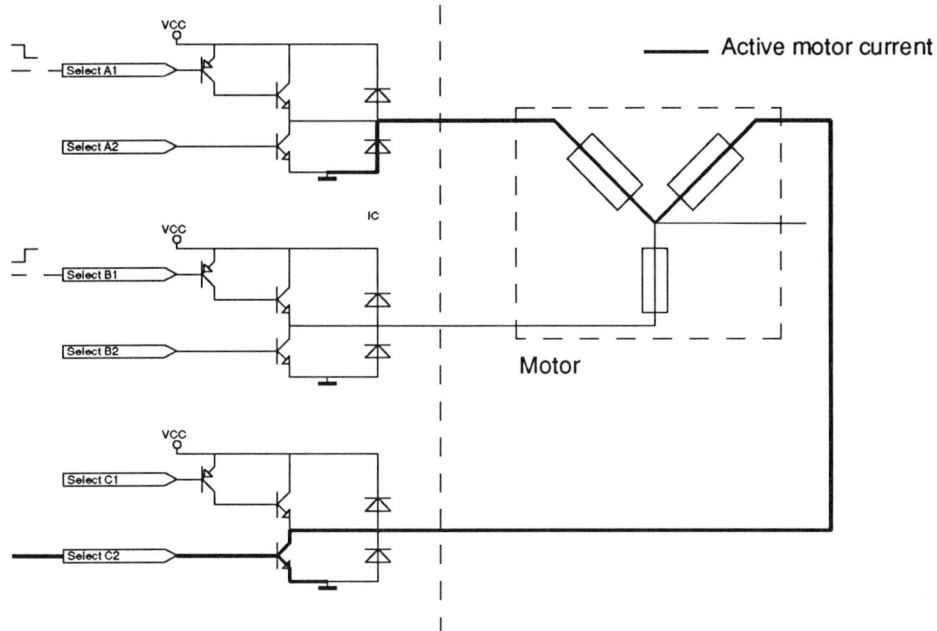

Figure 4.56: Use of flybackdiode D1 during switching

Figure 4.54 gives the function of the various components in the power stage during one part of the oscillation cycle. During this part current flows through the Darlington pseudo-pnp transistor T1 of power stage A into the motor. From the motor the current flows through transistor T2 of stage B. At a certain moment, determined by the commutation logic, the function of stage B is switched to stage C. This causes a reallocation of currents through the motorcoils with the transient effect of a fly-back pulse of the now deactivated motor coil B through flybackdiode D1 of stage B. See figure 4.55.

In a similar way the commutation logic switches (after a certain time) the function of power stage A to stage B. In this case the fly-back pulse of motor coil A flows through flybackdiode D2 of stage A. See figure 4.56. Transient signals are presented in figure 4.53.

From this diagram it is possible to derive most stressors for the components described in figure 4.54. Only for thermal considerations will it be necessary to look upon the entire circuit. As batch production and use of this integrated circuit will start in 1990 it is not yet possible to give detailed susceptibility limits. Therefore this section will concentrate on the description of relevant susceptibility limits and will not give detailed figures. First the failure mechanisms applicable to the integrated circuit as a whole will be discussed. In the succeeding sections the components described in the power stages of figure 4.52 will be discussed individually.

4.6.3.1 *Power dissipation and secondary diffusion*

To prevent problems with either the thermal-mechanic or the thermal-chemical behaviour of this circuit the manufacturer of this circuit defines a maximum junction (chip) temperature of 150 ^oC. To prevent excessive chip temperatures due to internal power dissipation the circuit contains a thermal protection circuit. See figure 4.52. As an important part of the power dissipation is caused by the three power stages the thermal protection circuit shuts down the operation of the power stages (= switches all power transistors to "open") in case the chip temperature exceeds a certain threshold temperature. First measurements of the manufacturer show a threshold temperature between 130 and 150 ^oC. This thermal protection only protects the circuit from the influence of excess dissipation in the power transistors and not from other (ambient) thermal influences.

4.6.3.2 *Current breakdown and electromigration*

As current breakdown and electromigration are related especially to areas within components where high current densities occur, the major bottlenecks for current breakdown appear to be the bondwires and/or metalization patterns with high current densities. Analysis on these failure mechanisms shows, however, no measurable influence of these failure mechanisms. Further analysis on larger batches for a longer time will be necessary for the development of a better defined susceptibility model.

4.6.3.3 *Avalanche breakdown*

Avalanche breakdown might occur in all components in the IC where reverse bias situations occur. Analysis on the first (four) samples show that avalanche breakdown occurs not under a reverse bias voltage of 21 V. This value is only usable as a first indication of a future more detailed susceptibility model. Further analysis on more samples will be necessary to obtain a statistical sound susceptibility model. Breakdown voltages were measured both across the parallel transistor/diode combination T1 and D1 and combination T2 and D2.

4.6.3.4 *Flyback diodes D1 and D2, pulse power effects*

The pulse power effects in the flyback diodes are for a large part determined by the external circuit. At the moment a power transistor switches off the energy stored in the corresponding motor coil will fly back. To prevent excessive voltage peaks diodes D1 and D2 are used as protection. A consequence of this method of protection is the fact that both diodes are stressed with pulses with steep slopes. Therefore one of the first susceptibility limits to test in practice is the susceptibility for power pulses, both during switching on- and switching off- of the diode. First tests indicate that in the measured pulses do not show the characteristic power peaks (see figures 4.19 to 4.21) related to fast switch-on and switch-off of diodes. See figures 4.57 to 4.59. A more detailed study of the susceptibility of the used diodes for this failure mechanism will be necessary at the moment that sufficient samples of this IC are available.

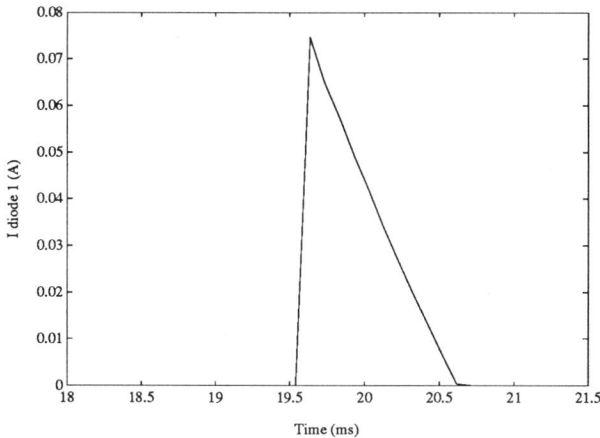

Figure 4.57: Diode current through flyback diode D1

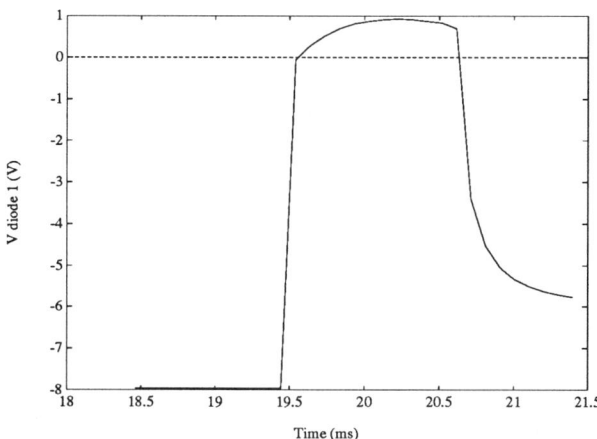

Figure 4.58: Diode voltage across flyback diode D1

4.6.3.5 *Switching transistors T1 and T2, second breakdown*

One of the major problems in acquiring susceptibility limits for the second breakdown failure mechanisms is the impossibility to have direct access to the base current of both transistor T1 and transistor T2. Both signals are internal

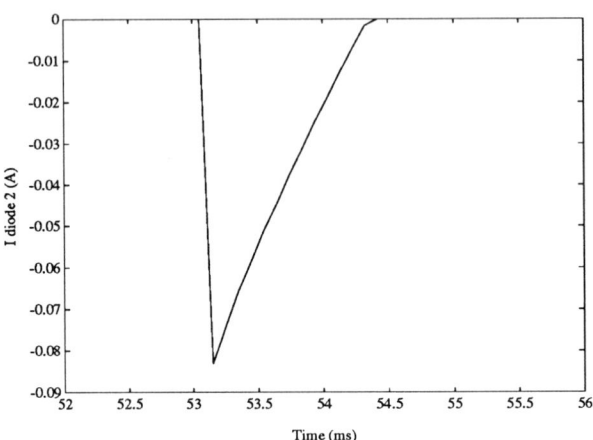

Figure 4.59: Diode current through flyback diode D2

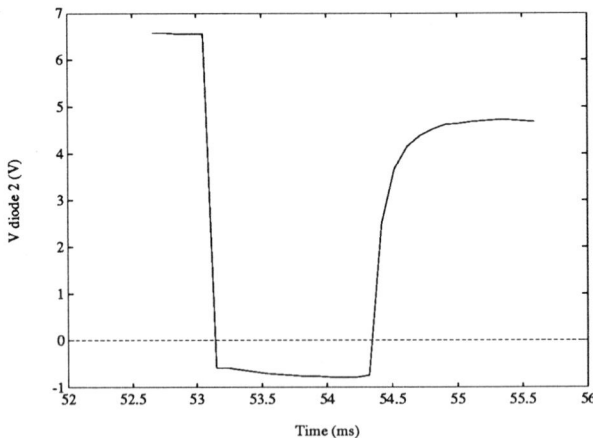

Figure 4.60: Diode voltage across flyback diode D2

signals in the integrated circuit. As both transistors are driven by the commutation logic of the integrated circuit it is possible to assume the exact characteristics of the slopes of the base currents as unknown but constant. This leaves the second breakdown as a function of the stressors V_{ce} and I_c of both transistors. First measurements have shown reverse second breakdown for V_{ce} values of about 18 V combined with collector currents of about 0.9 A. Further research on the susceptibility for this failure mechanism will be necessary at the moment that sufficient samples of this IC are available.

4.7. **Summary of susceptibility models**

This chapter has discussed some (often well-known) failure mechanisms occurring in electronic components. From these failure mechanisms susceptibility models were derived for a number of practical components. For components X and Y this resulted in a complete susceptibility model. Although the number of samples (four) of circuit Z was inadequate for the development of susceptibility models the only bottleneck is this number of samples; using the described technique it is also possible to develop susceptibility models for this category of integrated circuits.

In theory much of the information presented in susceptibility models should be available in the manufacturer's component databooks. Unfortunately this is quite often not the case. In many cases own experiments were necessary for the development of susceptibility models. Chapter 5 will show that components tend to fail on failure mechanisms of which quite often insufficient information is available.

As described in Chapter 3 it should be possible to use the discussed susceptibility models in the analysis of practical circuits in order to obtain more detailed information related to component failures in practical circuits. In order to demonstrate the use of susceptibility models, Chapter 5 will analyse some circuits using the presented susceptibility modes. This chapter will also relate practical failure figures to the stressors occurring in circuits and to the discussed susceptibility models.

5
Stressor Sets for Practical Circuits

5.1. Introduction

In the previous chapter models were derived to estimate the susceptibility of a component for certain (combinations of) stressors. Before it is possible to use these susceptibility models in a further analysis it will be necessary to determine such a stressor set for a given circuit. As was mentioned in Chapter 3 it is important that during the deriving of stressor sets two important aspects are taken into account. For one single circuit it is important that the stressor set of a component covers variations in stressors within this circuit. For a large batch of circuits it is important that the differences between the individual circuits are taken into account. This leads to the following statements:

> *Every individual stressor set should be ergodic* (5.1)

> *Every mean stressor set should represent the entire circuit batch* (5.2)

This means that the distribution of the stressor set should converge to the distribution of the signals related to these stressors for an infinite observation time. See figure 5.1. This implies that:

> *A stressor set for a component should cover all (quasi-) stationary states. (ad 5.1)* (5.3)

> *A stressor set for a component should cover all transitions between the various states. (ad 5.1)* (5.4)

> *A mean stressor set should be based on individual stressor sets. (ad 5.2)* (5.5)

> *The individual stressor sets should represent the stressors in one individual circuit. (ad 5.2)* (5.6)

> *The total of all individual stressor sets should reflect the behaviour of the complete batch.* (5.7)

This implies that there are two factors of great importance for the analysis of detailed stressor sets:

Variations in stressors due to variations in circuit behavior within   (5.8)
one single circuit.

Variations in stressors due to differences in circuit behavior   (5.9)
between circuits within a large batch of circuits.

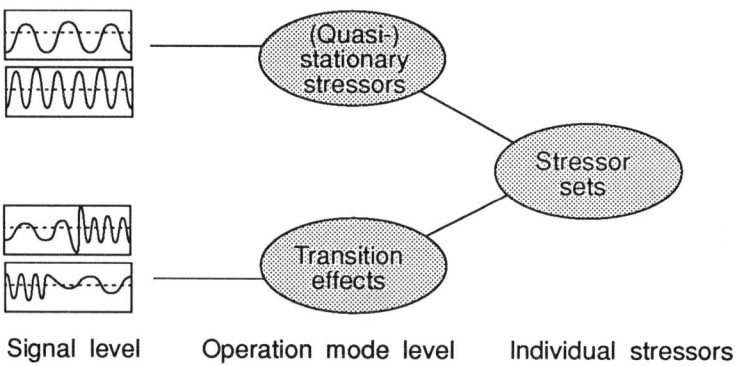

Figure 5.1: Stressor should cover states & transitions

It is possible to describe these influence factors using the hierarchical structure of figure 5.2.

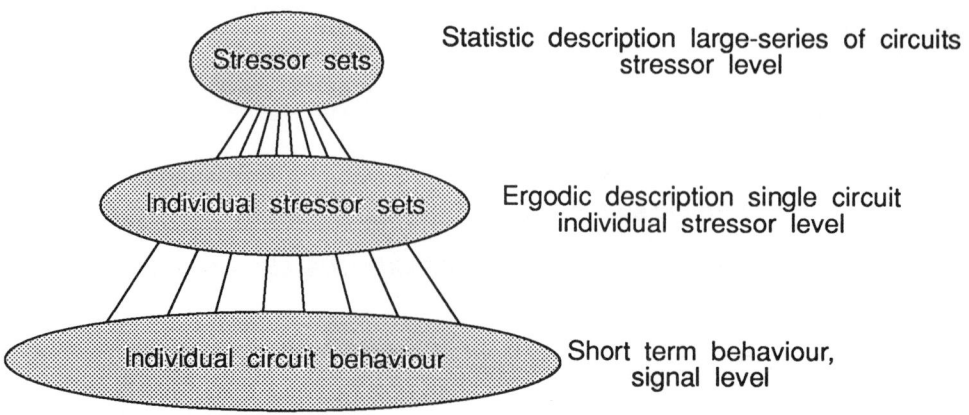

Figure 5.2: Hierarchy of stressors

This chapter will show how it is possible to derive stressor sets for practical circuits. For this purpose this chapter will derive stressor sets for two components, also discussed in chapter 4. These components are:

— Diode X (Schottky diode)

— Transistor Y (high voltage transistor)

Finally this chapter will use the stressor sets, obtained in this chapter, and the susceptibility of these components, derived in Chapter 4, to predict the failure behavior of these component in relation to the developed stressor/susceptibility models.

Before going into details of the acquisition process it is important to summarize what stressors, in terms of electrical signals, have to be taken into account.

Failure mechanism	Stressor	Related electrical parameters (simulation)
Current breakdown	Current density	I_{branch}
	Environmental temperature,	
Thermal cracks	Dissipated power	Power dissipation I_{branch}, V_{node}
	Environmental temperature,	
High-voltage breakdown		
(Impact ionization)	Electric field	V_{node}
(Avalanche and Zener)	Electric field	V_{node}
(Electron trap ionization)	Electric field	V_{node}
	Environmental temperature	
	Time dependent behaviour of the electric field: $\frac{dE}{dt}$	$d\frac{V_{node}}{dt}$
Corrosion (time dependent)	Environmental temperature (negative influence on susceptibility)	
	Dissipated power (negative influence on susceptibility)	Power dissipation I_{branch}, V_{node}
	D.C. voltage	V_{node}
	Moisture,	
Electromigration (time dependent)	Current density	I_{branch}
	Environmental temperature,	

Failure mechanism	Stressor	Related electrical parameters (simulation)
Secondary diffusion (time dependent)	Environmental temperature	
Switch-on effect (diodes)	Re-polarization speed {roman {d}V} over {roman {d} t }	$\dfrac{dV_{node}}{dt}$
	Speed of charging {roman {d}I} over {roman {d} t }	$d\dfrac{I_{branch}}{dt}$
Switch-off effect (diodes)	Stored charge Qs in the diode at the moment of polarity reversal	I_{branch} (before polarity reversal)
	Re-polarization speed {roman {d}V} over {roman {d} t }	$d\dfrac{V_{node}}{dt}$
	Maximum reverse diode current	I_{branch} (directly after polarity reversal)
	Applied reverse voltage	V_{node} (directly after polarity reversal)
Forward bias second breakdown (bipolar transistors)	Collector emitter voltage	V_{node}
	Slope of the base current during switching-on $\dfrac{dIb}{dt}$	$d\dfrac{I_{branch}}{dt}$
	Slope of the collector current during switching-on $\dfrac{dIc}{dt}$	$d\dfrac{I_{branch}}{dt}$
	Environmental temperature	
Reverse bias second breakdown (bipolar transistors)	Collector-emitter voltage	V_{node}
	Stored charge in the transistor collector (and base) current at the moment of transistor switch-off	I_{branch} (before polarity reversal)
	Discharge speed {roman {d}I} over {roman {d} t } positive influence: *fast discharging of the transistor base* negative influence: *too fast discharging of the transistor base will cause "charge bubbles".*	$d\dfrac{I_{branch}}{dt}$
	Environmental temperature	

For the backgrounds of this table: see Chapter 4. From this table it is possible to derive the following conclusions:

> *It is possible to relate most stressors to the branch currents and*   (5.10)
> *node voltages of a circuit*

> *Sometimes it will be necessary to use differentiation or integra-*   (5.11)
> *tion of these voltages and currents with respect to time*

In other words: it is possible to derive stressors from measurable parameters or from parameters for which calculation using a circuit simulation program is possible. The only difficulty in this respect is the environmental temperature. For a detailed stressor/susceptibility analysis it will be necessary to simulate not only the electrical behaviour of a circuit but also the thermal behaviour. In this book the environmental temperature will be treated, for reasons of simplicity, as constant.

The problem in the relation between stressors and voltages/currents is the fact that sometimes integration or differentiation of signals is necessary. This might cause problems, both in computer simulation and in practical measurements. Therefore if it is possible to relate stressors (if necessary indirectly) to signals where these mathematical operations are not necessary, the analysis of these signals should be preferred.

5.2. Acquiring stressor sets

Generally speaking there are two possible ways to obtain stressor sets. First there is the possibility to derive stressor sets from practical measurements. In those cases where sufficient systems are available it is possible to do a statistical evaluation of the individual stressor functions existing in individual systems. As the stressor sets are dependent on the conditions of use and the operation modes of a system it is important that the measured stressor is based on all possible operation modes of a circuit and all the possible transitions between the various operation modes. This can become quite a tedious job, especially as the entire operation has to be repeated for a number of systems to obtain an accurate statistical mean stressor model.

Therefore generation of stressor sets using computer simulations has tremendous advantages. In those cases where realistic computer simulation models for components exist it is possible to derive all circuit signals using one single simulation. Second there is the possibility to simulate the effect of parameter changes. This has many advantages over measuring the effects of parameter changes. In those cases where a realistic tolerance model of a component is available the changing of a component for another component consists of entering a new numeric parameter value. The following sections will discuss both acquisition methods:

— Computer simulation

— Actual measurements

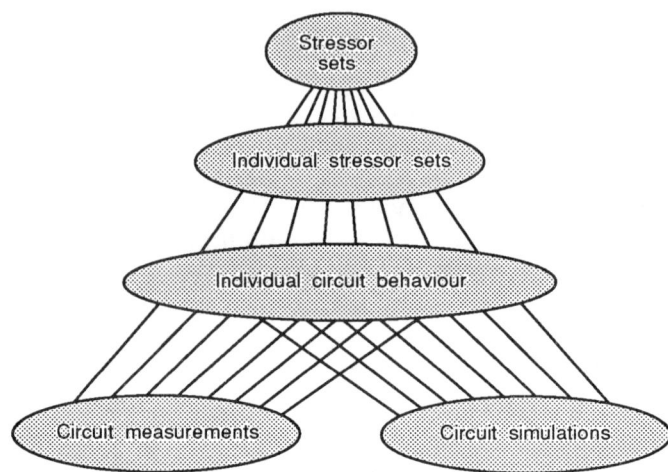

Figure 5.3: Stressors using measurements or simulation

See figure 5.3. The following sections will concentrate on the possibilities and the limitations of using these acquisition methods. The main item in these sections is how measurements and computer simulations relate to the demands for the acquisition of stressor sets.

5.3. Deriving stressor sets from computer simulation results

As mentioned earlier in this section the use of computer simulation as a method for stressor acquisition has considerable advantages. It is nevertheless important that computer simulation used to derive stressors fulfills a number of demands. These demands are closely related to the demands on stressors.

— Computer simulation should give a realistic representation of real circuits, especially in respect with stressors

— It should be possible to derive individual, ergodic stressor descriptions from the simulation results

— The mean stressor set, derived from the individual stressor sets, should reflect the behaviour of the entire batch of circuits

In other words: computer simulation to derive stressor sets should be able to take into account differences within a circuit due to the different operation modes of a circuit and differences between circuits due to differences in individual components. These demands relate to a number of properties in respect to circuit simulation. For the acquisition of stressor sets using computer simulation it is necessary to have the following items available:

— Realistic functional component models

— Realistic component tolerance models

A third demand, also of great practical significance, is the availability of

— Simulation software, able to use the models mentioned above

5.3.1. *Requirements for simulation software*

To start with the last demand: practical tests using available simulation software* showed that although most software packages are able to simulate the functional behavior of many circuits quite accurate, the acquisition of stressor sets results in some additional demands on accuracy. This will be illustrated using a simple example. One of the failure mechansisms, discussed in Chapter 4, is the mechanism of reverse bias second breakdown, occurring in bipolar junction transistors. For many functional simulations it is sufficient to simulate, for example, the node voltages and branch currents. The derivation of stressor sets requires additional information. One of the important stressors for this failure mechanism is:

$$\frac{dI_b/dt}{I_c} \tag{5.12}$$

First conclusion from this expression is the necessity to use a computer simulation program to calculate the time functions of the base- and collector-currents for this transistor. Due to the non-linearity of base- and collector current it is necessary to derive the dI_b/dt using numerical differentiation. This numerical differentiaton can be a source of considerable inaccuracies. See figure 5.4 for a practical example of collector- and base- currents derived from simulation. For many functional applications the accuracy of this simulation is quite satisfactory. Detailed analysis of the actual time steps used in the simulation process learns that, in the given example accurate calculation of the dI_b/dt is impossible, due to inaccuracy of the results of this differentiation. See figure 5.5.

This leads to the following result:

> *The required accuracy (in time and in value) of the results of* (5.13)
> *computer simulation, used to obtain stressor sets, is related to*
> *the nature of the required stressor sets.*

---

*   The simulation packages Spice (Berkeley Spice, Espice, Hspice, Pspice), Philpac and Microcap have been used

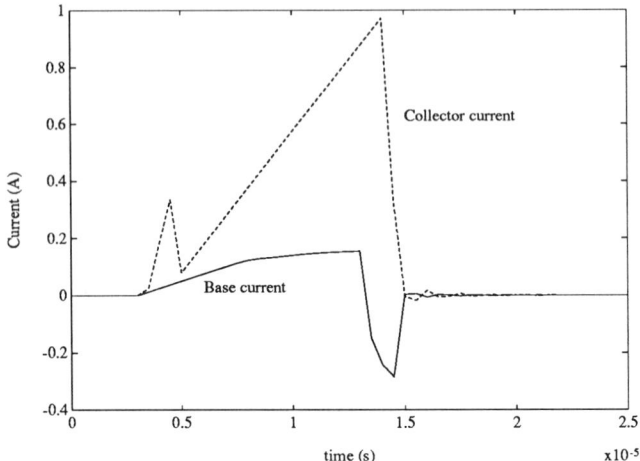

Figure 5.4: Simulation results used for stressors

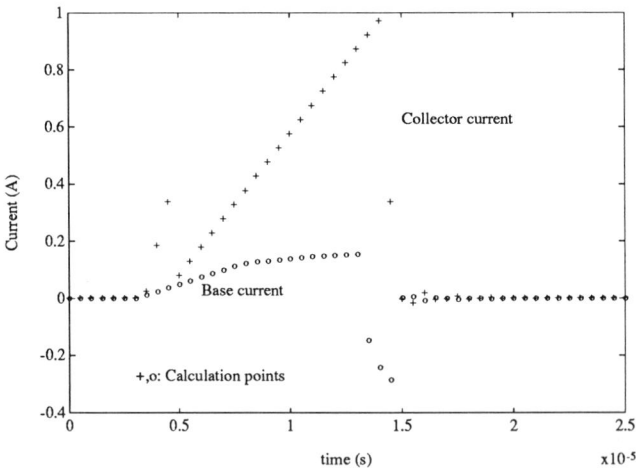

Figure 5.5: Accuracy problems relevant for stressors

This is quite often a problem in practical situations. Earlier experiments during this project resulted sometimes in the erroneous presentation of stressor sets, not based on circuit behaviour but on numerical problems with the results of the simulation software.

— There are many circuit simulators available for the simulation of the signals required in the determination of stressor sets.

— Major difficulty in this respect is the accuracy (in time and value) of the simulation results, especially where numerical differentiation is involved.

5.3.2. *Requirements on functional component models*

One of the major constraints for the usability of simulation software is the availability of adequate models, describing the behaviour of real-world components. For simple components, such as resistors, it is quite often adequate to use a simple one parameter model. For more complex components, such as semiconductors, quite often many models are available differing in the amount of detailed effects described. A general guideline for the use of component models as part of the deriving of stressor sets is to use the most detailed available model. Due to the amount of detail (see also the previous section) required for the deriving of accurate stressor sets the more simplified transistor models tend to cover only partially the effects related to failure mechanisms. This can be illustrated using the example of the bipolar transistor*.

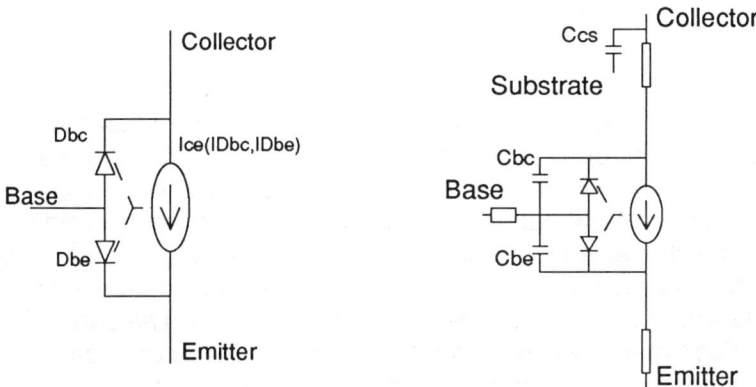

Figure 5.6: Ebers-Moll 1   Figure 5.7: Ebers-Moll 2 model

The SPICE simulation program is able to use three different bipolar transistor simulation models, each different in the amount of detailed effects described: Ebers-Moll I, Ebers-Moll II and enhanced Gummel-Poon [Get78]. See figures 5.6, 5.7 and 5.8 for the details of the simulation models. Of these models only the Gummel-Poon model is able to simulate effects related to a non-homogeneous base structure. This is relevant, especially for simulation of stressors related to the second breakdown failure mechanism.

---

*   Although this section uses bipolar transistors as example the presented problems do
     also occur (in a modified form) for other components

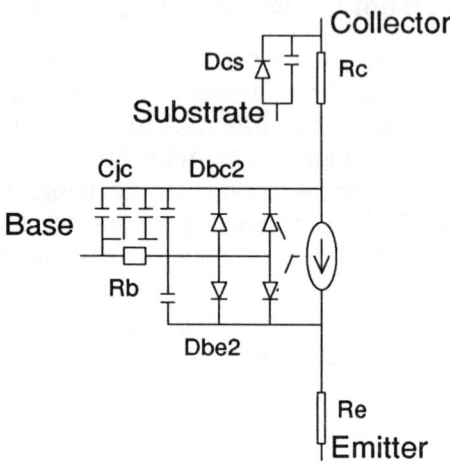

Figure 5.8: Gummel-Poon transistor model

The disadvantage is that in many cases even the most detailed simulation models available do not provide a sufficient level of detail. This problem especially rises in the cases where additional effects, not described in the current available models have to be taken into account. An important example in this respect is the functional effect of current crowding or pinch-in (described in Chapter 4) which is closely related to the stressor sets relevant for failure mechanisms like forward- and reverse- second breakdown. The currently available Gummel-Poon model is able to cover some effects of a non- homogeneous base. It is not able to cover geometrical effects, relevant for pinch-in or current crowding as it assumes a more or less circular transistor. The effects, mentioned above, are closely related to the geometry of the transistor. This geometry is in most cases not circular. Using the finite element option, described in Chapter 4, is one of the ways to cover parts of the transistor geometry but it does not result in an open, easy to use, simulation model.

To resume, this leads to the conclusions that for many components computer simulation models are available. Applying these models requires precautions that not a model lacking the required detail is used. For the analysis of some failure mechanisms more detailed functional models are required (geometrical effects of second breakdown).

5.3.3. *Parameters, required for simulation models*

Important aspects of simulation of components are not only the model but also the model parameters. Nowadays many component manufacturers (or third-party organizations) are able to provide model parameters for many practical components.

In those cases where simulation models of components are not available there is also the possibility to derive a simulation model from practical measurements. Getreu [Get78] presents, for example, guidelines to derive simulation parameters for bipolar transistors from practical measurements. The disadvantage of this method is the fact that only one transistor is used in the modelling process. How well this single transistor represents the "typical" behaviour is in this respect an important question.

A third method to derive model parameters is to use the component manufacturers databooks. The intention of these databooks is, amongst others, to present functional parameters and graphs characterizing the behaviour of the presented product. Databooks are intended especially for use in the design process of systems to provide the user of components with adequate component data. During this project quite often the situation occurred that although quite detailed information was presented in the various databooks, important information relevant for stressor/susceptibility models was missing. Therefore three sources of data were used to derive component parameters:

— Manufacturers (or third party) complete simulation parameters

— Parameters derived from own measurements

— Data obtained from manufacturer's databooks

In those cases where the first option was available this option was used (under the condition that all relevant parameters were present). In those cases where the databooks provided sufficient data the third option was used and if both methods were not able to provide data, own models were developed. See also Appendix C. One of the major problems in all these methods is that nearly always the simulation parameters reflect the behaviour of one single nominal component. To take the effects of individual differences between components into account it is necessary to derive not only simulation parameters but also to introduce the effects of parameter tolerances.

5.3.4. *Requirements for component tolerance models*

On of the aspects of stressor/susceptibility analysis is obtaining information (in terms of stressors) reflecting the differences between the individual circuits. One of the most significant causes for these differences is related to the differences in individual components. Every circuit is built using individual (= different) components and will therefore show individual differences in behaviour. The important conclusion of Chapter 3 was the necessity to distinguish differences between individual components on the level of individual circuits. For simulation of circuits this means that not only component models (with the related component parameters) should be available but also component tolerance models with the related parameter tolerances. Another important aspect in this respect is the *correlation* between the various parameters within a component. A component tolerance model describes the nature of parameter variations, the related parameter distribution and the mutual dependency of the different parameters within a component. With the use of such a component tolerance model it is possible to give realistic predictions of the probability of occurrence of components having certain combinations of parameters. See also figure 5.9.

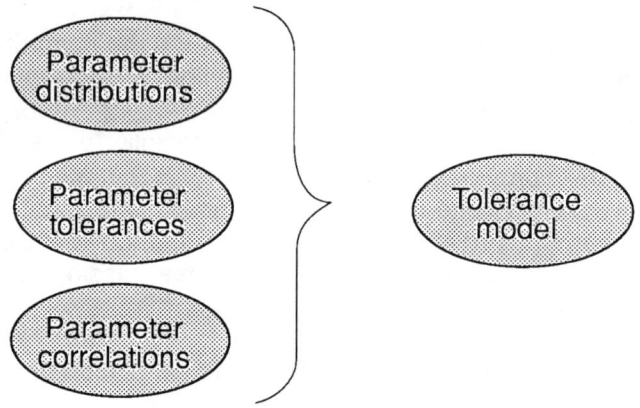

Figure 5.9: Tolerance modelling

Unfortunately for many components tolerance models do not exist, especially where components have more than one parameter.

Related to the sources of data mentioned in the previous section there are three possible sources of data which might be of importance regarding tolerance-models and data:

— Manufacturers (or third party) complete tolerance models

— Parameters derived from own measurements

— Data obtained from manufacturer's databooks

Although many books have been written on models of electronic components there is only very little information known about the tolerances in the parameters of simulation models or correlation between these parameters. Manufacturer's databooks either present nominal parameter values or nominal values and maximum- and minimum- limits. Distributions of parameters and correlations between parameters are not given.

The additional problems of component tolerance models in computer simulations are caused by the simulation software packages. Many simulators are not able to use other parameter distributions than uniform. In many packages it is not possible to describe correlations between parameters. See also Appendix C. The practical problems of the simulation packages themselves are not discussed in further detail in this book.

5.3.5. *Summary of the demands on circuit simulation for stressor/susceptibility analysis*

Circuit simulation provides means and methods to simulate electrical parameters relevant for the development of stressor sets. It is important that precise simulation- models and parameters are used. The available simulation models are in some cases not adequate to simulate effects relevant for failure mechanisms such as second breakdown, but a more important problem is the unavailability of valid tolerance models and valid parameter tolerances.

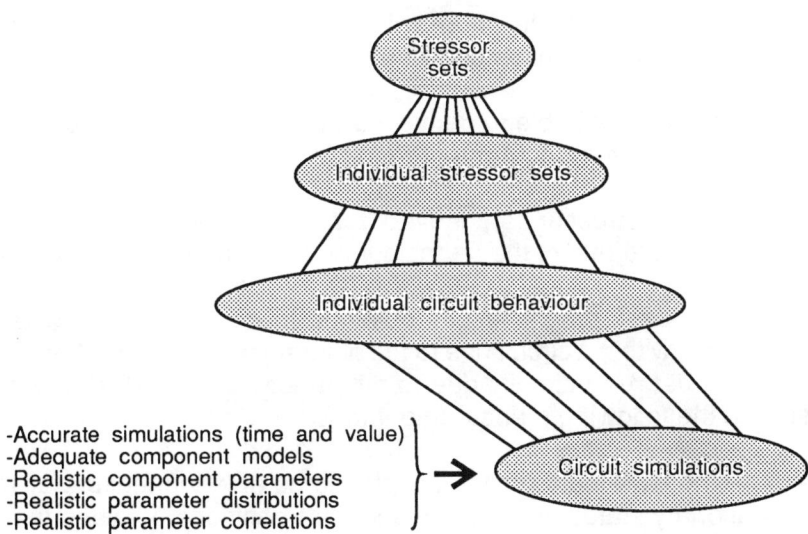

Figure 5.10: Requirements simulation, stressor aquisit.

This is a serious drawback for stressor/susceptibility analysis. As mentioned, especially in Chapter 3, one of the main sources of reliability problems can be the difference, either in functional parameters, or in susceptibility, of the different components within a batch. This is also one of the most important tests to verify the presented stressor/susceptibility method. Analysis of practical failures should relate failure causes to stressor/susceptibility interaction.

To allow computer simulation in stressor/susceptibility analysis tolerance models were developed for some practical components. See also appendix C of this book. Due to the unavailability of some detailed functional models but especially due to the unavailability of tolerance models it is at this moment not (yet) possible to use simulation as a yielding way to acquire stressor sets. The following chapter will show that in those cases where these models are available they result in a significant improvement in reliability analysis and reliability optimization compared to traditional methods.

5.4. Deriving stressor sets from practical measurements

As mentioned in the introduction of this chapter deriving stressor sets from measurements is a possible alternative to simulation. The demands for deriving stressor sets from measurements are similar to the demands for deriving them from simulation data:

— Measurement results should give a realistic representation of stressors occurring within a circuit

— It should be possible to derive individual, ergodic stressor descriptions from the measurement results

— The mean stressor set, derived from multiple individual stressor sets, should reflect the behaviour of the entire batch of circuits

In other words: Measurement results obtained to derive stressor sets should reflect differences within a circuit due to the different operation modes of a circuit and differences between circuits due to differences in individual components. This results in the following measurement requirements:

— The measurement equipment should be able to measure all (quasi) stationary states of a system and the transitions between the various states.

— An adequate sample series of circuits should be available, reflecting the behaviour of the entire circuit population.

The following sections will convert these demands into practical guidelines how to derive stressor sets from practical measurements.

5.4.1. *Requirements for measurement hardware*

It is possible to derive an important part of data required for the deriving of stressor sets from the transient voltage- and current- signals in a circuit. One of the simplest ways to acquire these data is to sample voltages and currents, to convert the data to a digital form and to derive the stressor sets from this digital data. There are two requirements in relation to data sampling:

— Accurate description of a stressor set will require a sampling frequency greater than two times the highest frequency in the stressors frequency spectrum.

— Accurate description of a stressor set will require a number of samples sufficient to cover all the different states of a system.

The result of the first demand will often be a high sample frequency while the second demand will result in a large elapsed sample time. In practical situations these combined demands can result in quite unacceptable numbers of samples. A sample frequency of 30 MHz and an observation time of 24 hours would result in 2.59×10^{12} samples.

As a signal has often one or more independent quasi-stationary states, each characterized by their stressor set, it is possible to derive the overall stressor set function from the individual state stressor sets using:

$$f_{\text{str},y}(x) = \sum_{i=1}^{n} \frac{T_i}{T_{\text{total}}} f_{\text{str},y,i}(x)$$

(3.18)

where $f_{str,y,i}(x)$: the stressor probability density function of quasi-stationary state i.

$\dfrac{T_i}{T_{\text{total}}}$: the fraction of time that the stressor is in quasi-stationary state i.

See figure 5.11. This gives the following demands for acquisition of stressor sets:

— Accurate description of a stressor set will require a sampling frequency greater than the highest frequency in the stressors frequency spectrum

— The total sample time of every sample serie should cover a complete (quasi-) stationary state or state transition of a system

— The total set of sample series of a circuit should result in an ergodic description of the stressor set

See also Chapter 3.

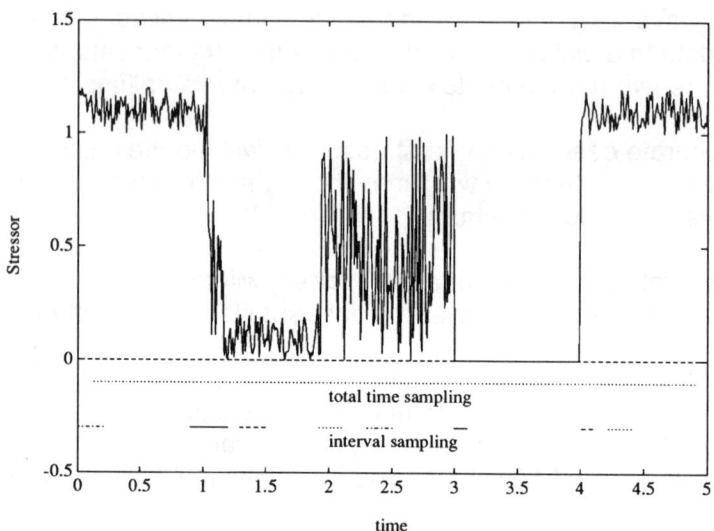

Figure 5.11: State stressors -> Individual stressor set

5.4.2. *Measurement of individual stressor-sets*

With the measurement equipment, able to fulfil the demands of the previous paragraph it is possible to derive stressor sets for most components within a circuit. For the measurement of stressor sets the following procedure is used:

Step	Action
1	Select component to be measured
2	Derive measurement demands (Voltages, current, slopes)
3	Adapt measurement system to measurement demands (mind time-differentiation)
4	Derive a table of all possible (quasi-) stationary states of a system
5	Derive a table of all possible state transitions of a system
6	Measure state-stressors and state transistion stressors
7	Use 3.18 to form a complete individual stressor set
8	Form joint stressor probability density functions for the entire stressor set

The last sections of this chapter will show practical stressor measurements. In this way it is possible to use this approach to measure many stressors for many components.

5.4.3. *Measurements of mean stressor sets*

One of the most important problems of acquiring mean stressor sets is how to get a usable impression of the stressor-sets in the entire circuit population. In a circuit with about 100 components, each with many parameters, it is very difficult to analyse all the effects of all possible combinations of component parameters on the stressor set of a given component. The first difficulty is to obtain components with certain parameters. Changing component parameters to other values during computer simulation is often a question of changing a number. Obtaining a component with certain (often extreme) parameters in a practical situation will quite often be a difficult job. Second problem is the enormous number of possible parameter combinations possible in one single circuit. See Chapter 6 for a practical example. Summarizing:

— The development of stressor sets using deterministic exploration of all possible parameter combinations is practically impossible.

Therefore it is necessary to use another approach. One of the main conclusions of Chapter 3 was that practical failures can be related to extremes in parameter values. In case of extremes in functional parameters, this results in extremes in the stressor function. In case of extremes in susceptibility parameters this results in failures under otherwise allowable stressors. As a consequence it should be possible to derive extremes of the various distributions from the analysis of circuits, failing under practical circumstances. During this project feedback of failing circuits was used as a part of the process of obtaining mean stressor sets.

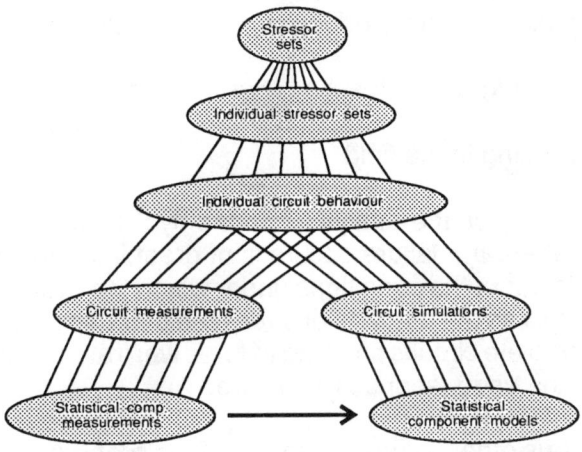

Figure 5.12: Measurements used to obtain models

The feedback of circuits was used not only for the acquisition of stressor sets. The circuits were also used to select components where the parameters had a dominant influence on the stressor/susceptibility interaction of (other) components. These dominant components were analysed in detail in order to derive detailed functional- and tolerance- models usable in future stressor simulation. See figure 5.12. In this way the experience, build up using actual measurements, is consolidated in simulation models. This allows the establishment of a widely accessible "knowedge base" in which, implicitly, knowledge about causes of earlier reliability problems is stored and maintained. See also Chapter 7.

5.4.3.1   *The use of pre-selected circuits*

To develop and verify stressor/susceptibility models it is necessary to have sufficient (failed) systems available for analysis. As mentioned in Chapter 2 one of the possibilities to acquire sufficient data is to take a limited number of systems and to use increased temperature as a method for accelerated testing. This solution however, is not usable in this particular situation. One of the main purposes of this research project is to find detailed failure causes occurring in practical situations and not to accelerate one known failure mechanism. Using thermal acceleration tests as a source of feedback data may result in accelerating a failure mechanism to a dominant effect on a higher temperature while the failure mechanisms at the temperature of normal circuit use is not dominant at all.

Therefore this project used the feedback of circuits failing under normal operation conditions. For feedback purposes the following failure sources were used:

— Circuits failing in the production during functional tests

— Circuits failing during two hour burn-in test

— Circuits failing in the field.

Especially for this project, the analysis of not only the failing component but of the entire circuit where the failure occurs is of great importance. Also as many details as possible about the conditions under which the failure occurred should be given. This simplifies the analysis of stressor distributions in problem circuits. Therefore only complete circuits showing failures were analysed. See figure 5.13 for a flowchart of the used feedback structure.

Circuits with obvious other failure causes, such as manufacturing failures (bad soldering) or system misuse were not used in the analysis.

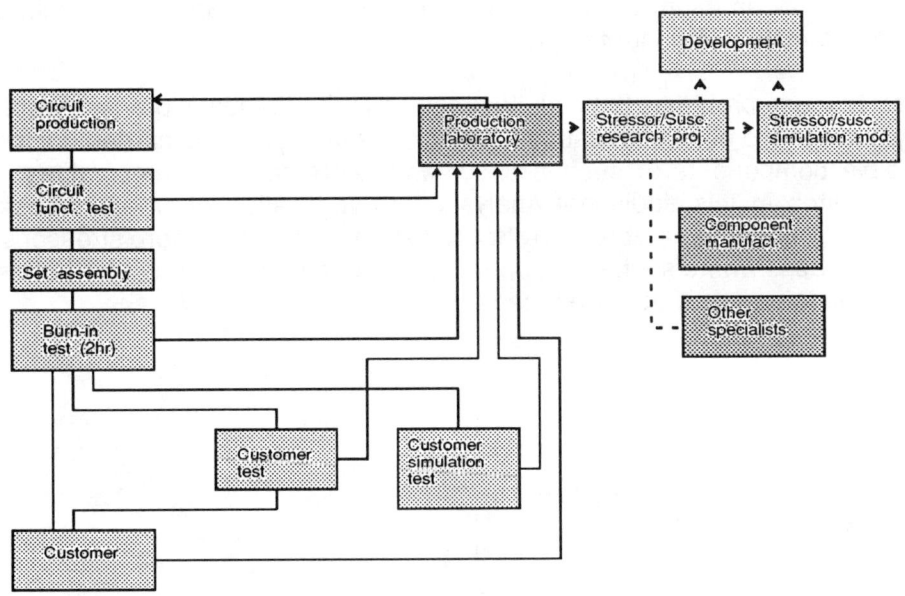

Figure 5.13: Feedback of failed circuit

5.4.3.2 *Relating failures in feedback circuits to mean stressor sets*

One of the main purposes of the described feedback process of failed circuits is the derivation of mean stressor sets on certain components. There are some considerable disadvantages in this method:

— In those cases where components in a circuit are destroyed it is hard to tell whether they were extreme before failure, either in function or in susceptibility.

— Only in those cases where other components than the failed component had a dominant influence on the stressor set it was possible to derive, post- mortem, a relevant stressor set.

Therefore the feedback mechanism, described in figure 5.13, is not sufficient for a complete analysis. In those cases where components in circuits have failed it is necessary to do additional research on parameter tolerances and component susceptibilities of these failed components.

Another problem of the feedback method is the fact that in many cases extremes in component populations are selected using this feedback process but the nature of the tolerance model remains unknown. Components with extreme functional values might be related to an individual sub-population or might be part of the main component population.

To solve these problems the circuits obtained in the feedback process were not only analysed to obtain stressor sets. Additional analysis was carried out to find possible components in such a circuit with extreme parameter values. The second step in this additional analysis was verification of whether possible extreme component parameters relate to extremes in one or more stressor sets. In those cases where such a component had a dominant influence on a stressor a detailed tolerance model was made.

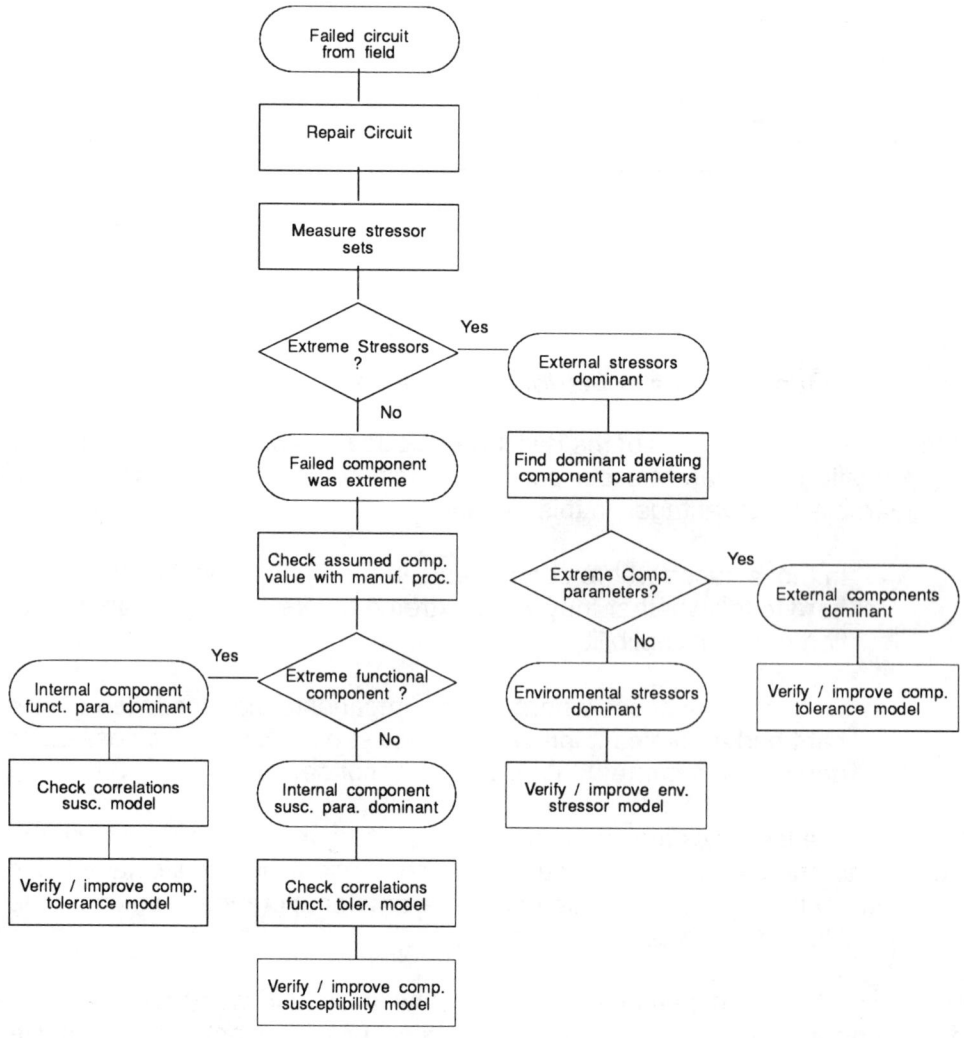

Figure 5.14: Development of models based on feedback

Using this approach it is not necessary to have a complete deterministic analysis of all possible parameter combinations within a circuit. Only in those cases where a clear relation between the failure probability of one component and the parameter values of another component exists the latter components are modelled in detail. See figure 5.14.

Summarizing in the analysis of failed (feedback) circuits the following steps were used:

Step	Action
1	Repair feedback circuit (replace failed components)
2	Check individual stressor set of circuit under test
3	In case of deviations in the individual stressor set: find components with dominant influences and check (or extend) tolerance models of these components
4	In case of no deviations in the individual stressor set: Try to relate the failure cause to functional/susceptibility distribution of the failed component

As a result the following components were analysed, either on tolerance model or on susceptibility model (or both):

— All components showing failures in practical situations are analysed to obtain a tolerance model and a susceptibility model

— All components with a dominant influence on the stressor set of a component failing in practical applications are analysed to obtain a tolerance model

In those cases where it was not possible to determine a realistic failure cause a number of verification steps was carried out.

— First the stressor set was checked on ergodicity. (Are there, for example, operation modes or transitions of the circuit not taken into account)

— Second the direct environment of the circuit was checked on environmental influence factors (electrical, thermal, etc.) having a possible relation to stressor sets occurring in the circuit

The following sections will present some examples of the acquisition of practical individual and mean stressor sets.

5.5. Practical stressor/susceptibility interactions

This section will show three examples of stressor acquisition for two practical circuits. Two practical components are discussed in these circuits:

— Diode X Schottky diode (circuit A)

— Transistor Y high-voltage transistor (circuit A and B)

This section will concentrate primary on the acquisition of stressor sets and the derivation stressor sets from practical feedback. The second purpose of this section is to show the use of feedback circuits in the process of modelling tolerance of practical components. Finally the results of the experiments are compared to the actual reliability figures of these components within their circuits.

As the stressors in a circuit are closely related to the operation modes of a circuit the table below gives the operation modes of the discussed circuits. As both circuits are power supplies in VCRs the operation modes of the power supplies are closely related to the operation modes of the VCR.

From / To	Power off	Stand-by	Play	Scan fwd	Scan rev	Wind	Re-wind	Pause	Rec-ord
Power off	S	T	T	T	T	T	T	T	
Stand-by	T'	S	T			T	T		T
Play	T'	T	S	T	T			T	
Scan fwd	T'	T	T	S	T			T	
Scan rev	T'	T	T	T	S			T	
Wind	T'	T				S	T		
Re-wind	T'	T				T	S		
Pause	T'	T	T	T	T			S	T
Rec-ord	T'	T						T	S

where   S: Stationary state
                                   T: Transition
                                   T': Transition (only in case of power interrupts)

This table is intended to give an indication of possible state transitions within a VCR. Depending on the actual type, changes in this table are possible.

5.5.1. *Diode X, circuit A*

As discussed in Chapter 2 and Appendix A the main purpose of diode X in circuit A is rectifying the secondary transformer voltage. See figure 5.15. This component is analysed in detail because of the considerable differences between the predicted failure rate (using traditional reliability prediction handbooks) and the actual failure rate.

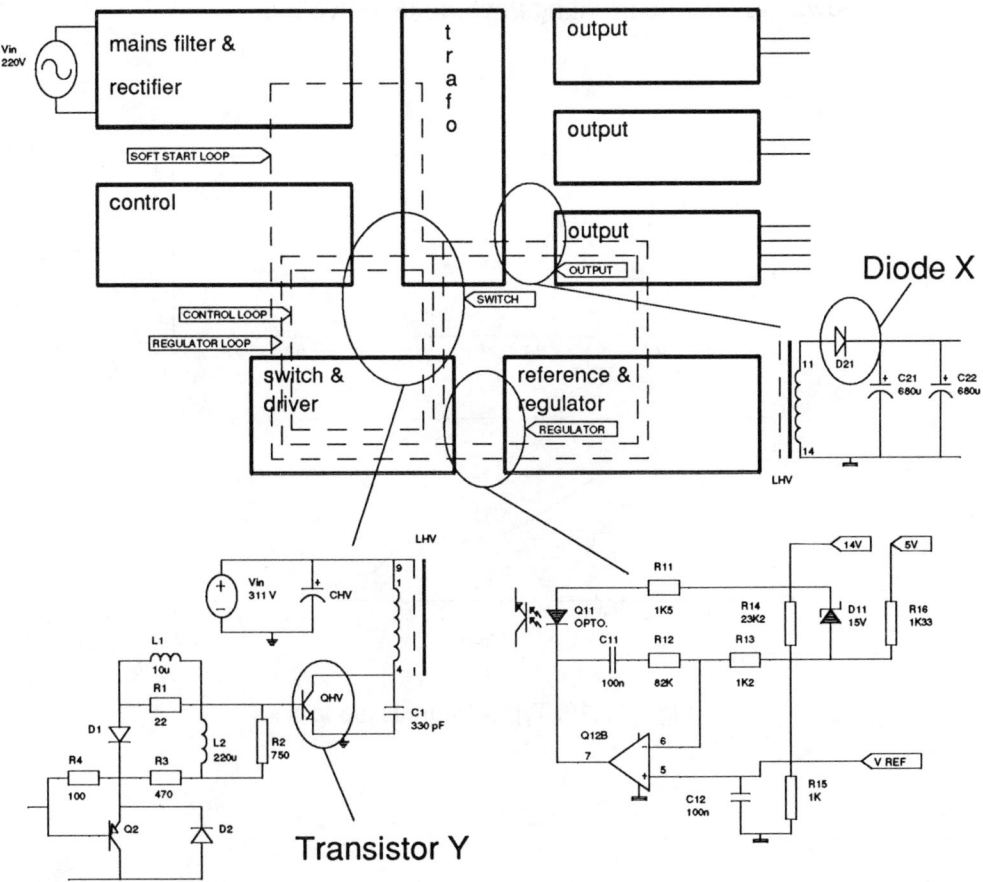

Figure 5.15: Components X and Y in circuit A

As mentioned in Chapter 4 the susceptibility limits for this diode are:

Parameter	Susceptibility limit
V reverse	V breakdown (see distribution break-down behavior)
T crystal	115 °C
{roman {d}V} over {roman {d} t }	$10^9 V/s$
{roman {d}I} over {roman {d} t }	$0.5*10^9 A/s$

Figure 5.16 shows signals on this diode. It is possible to derive that pulse-power effects do not occur in this application*. Remaining failure mechanisms are reverse breakdown and excess crystal temperature. It is possible to show stressors for both failure mechanisms in one diagram. See figure 5.17. This diagram shows the diode current against the diode voltage.

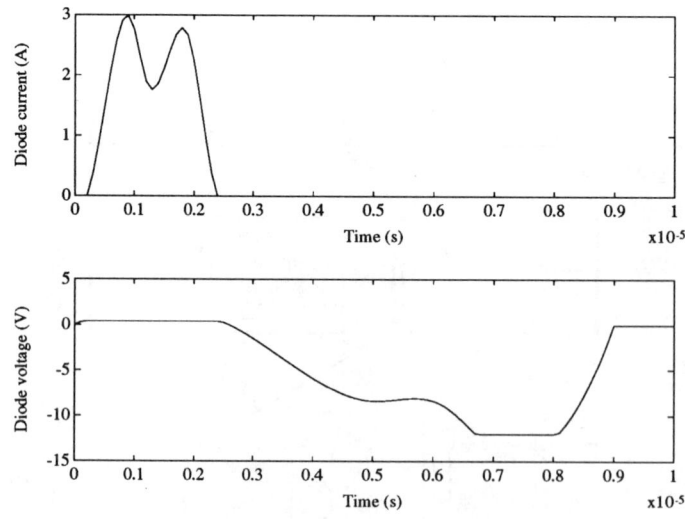

Figure 5.16: Typical signals diode X

---

*   More detailed analysis on pulse-power effects did not show relevant influence of this
     failure mechanism for this diode. Therefore pulse- power effects are not further
     discussed in this chapter.

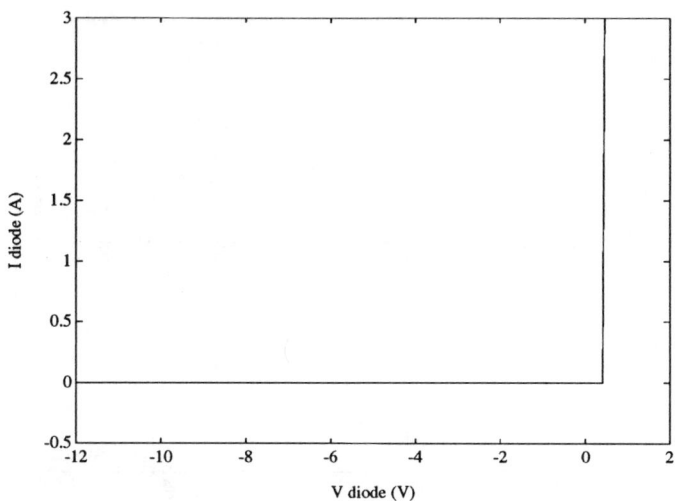

Figure 5.17: Typical voltage/current diagram for diode X

5.5.1.1   *Individual stressor set*

Using the measurement equipment, described earlier in this chapter, it was possible to derive a complete individual stressor set for the failure mechanism of this diode. See figures 5.18 and 5.19.

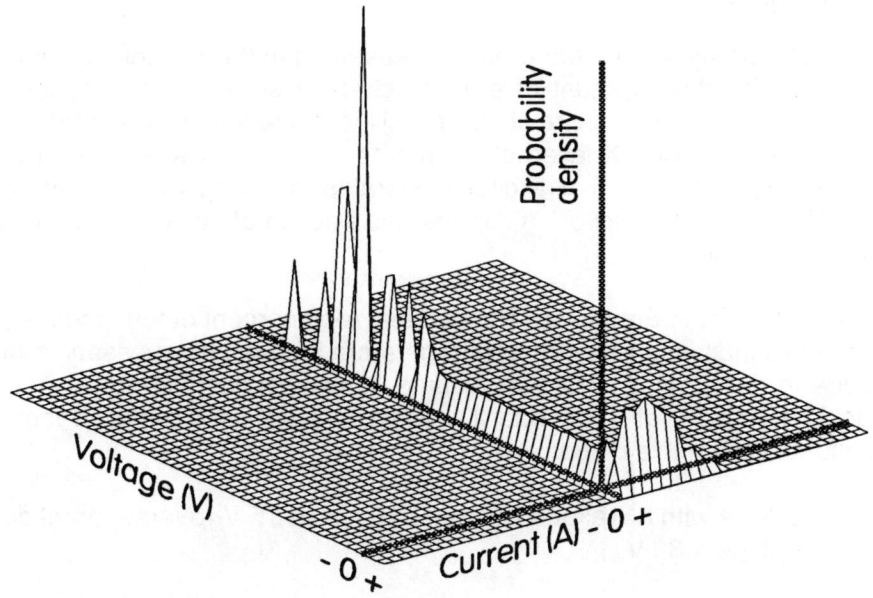

Figure 5.18: Joint individual stressor density diagram diode X

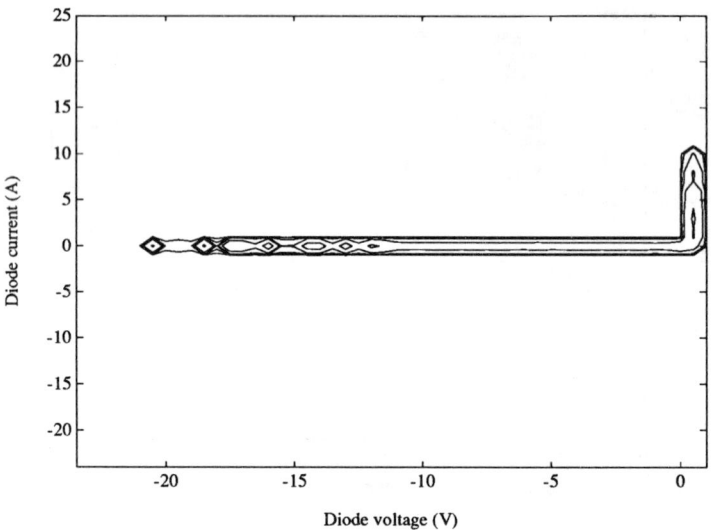

Figure 5.19: Joint individual iso stressor density diagram diode X

Figure 5.18 shows the stressor set with voltage and current as *x* and *y* axis and the probability density function as *z* axis. Figure 5.19 shows the same information, but now two-dimensional with iso-probability density lines connecting points with the same probability density.

5.5.1.2 *Mean stressor set*

According to the measurement protocol, presented in the previous sections the next step in the development of a mean stressor set is the use of feedback circuits. Actual feedback of circuits from Philips in Vienna showed that in about 1% of the circuits diode X failed at the moment the circuit was used or shortly thereafter. The first time that a circuit operates is during the functional tests directly after circuit production. A comparable fraction of this diode failed during the first year in the field.

Analysis of circuits in Eindhoven showed that replacement of the diode resulted in acceptable individual stressor sets. Therefore it became necessary to look to the diode in more detail. First step was to develop a complete tolerance- and susceptibility- model of this diode. Detailed analysis shows that in fact three populations exist of this diode:

— Diodes with a leakage current < 10 mA at 15 V reverse, breakdown voltage > 30 V

— Diodes with a leakage current > 10 mA at 15 V reverse, breakdown voltage > 30 V

— Diodes with a breakdown voltage < 30 V (see also Chapter 4)

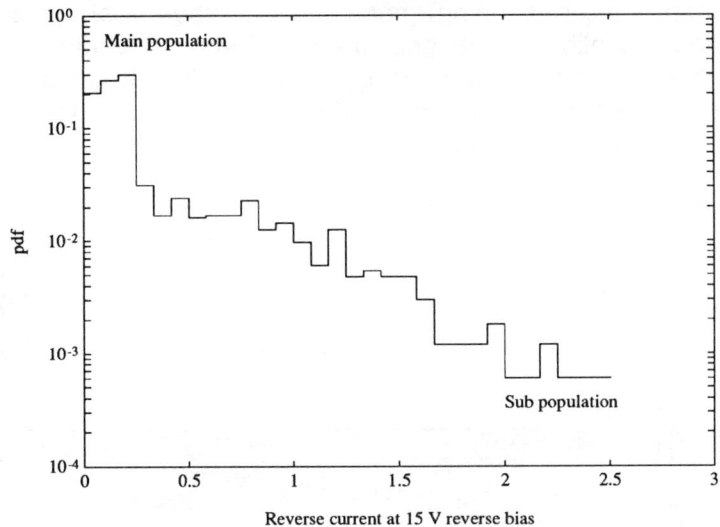

Figure 5.20: Leakage current at 15 V reverse bias

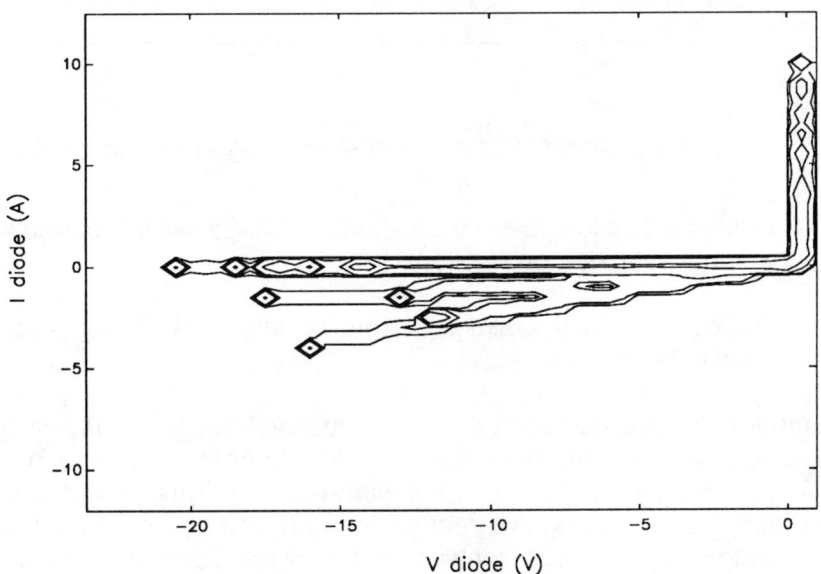

Figure 5.21: Joint mean stressor diagram diode X

It was not possible to use the third group (and a part of the second group) in measurements for a mean stressor-set as all diodes in the third group and some diodes in the second group failed immediately. See figure 5.21 for the remaining mean stressor diagram.

5.5.1.3 *Stressor/susceptibility interaction*

Figure 5.22 shows the (remaining) mean stressor diagram of this diode, combined with the susceptibility limits described in Chapter 4. Verification of actual failures at Philips Components Investigation Reliability Group (CIRG) shows indeed reverse power breakdown for this diode. See figures 5.23 and 5.24. From figure 5.22 it is possible to draw the following conclusions:

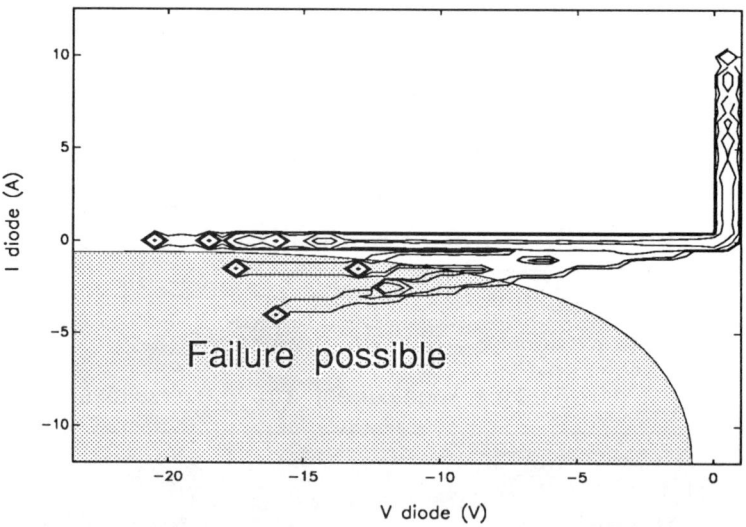

Figure 5.22: Stressor and susceptibility diagram for diode X

— Diodes with a low reverse breakdown voltage will fail immediately (during tests after production)

— Diodes with a high leakage current will fail, either during production-tests or later in the field

The number of failures occurring during production tests (~1%) is similar to the number of diodes with very high leakage currents or low breakdown voltages (~0.9%). The moment of failure of the high-leakage current diodes depends strongly on the exact leakage current and the applied environmental temperature. Laboratory experiments showed that a diode population exists where normal operation under normal ambient conditions is possible but increasing the ambient temperature will cause component failure. This makes the probability of

Figure 5.23: Diode X, failed due to reverse power breakdown

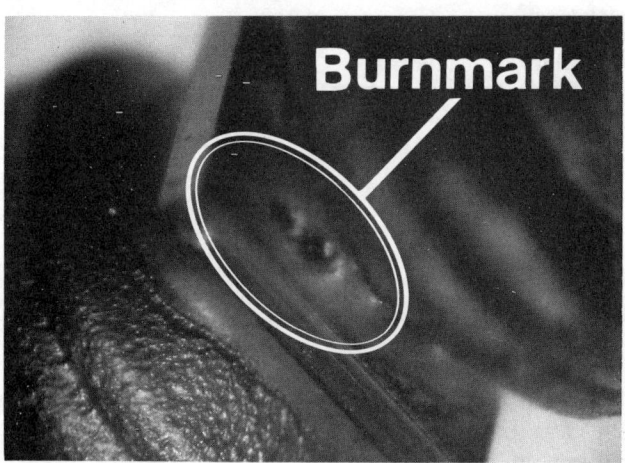

Figure 5.24: Diode X (enlarged)

occurrence of failure of such a diode strongly dependent on the long-term time characteristics of the environmental temperature.

Summarizing this leads to the following conclusions:

— Using stressor/susceptibility interactions it is possible to explain a considerable number of failures during production

— In this case stressor/susceptibility interaction relates the occurrence of certain failures to the occurrence of certain (high) environmental temperatures. Our own experiments have confirmed this interaction.

5.5.2. *Transistor Y in circuits A and B*

This section will discuss the stressor-sets of transistor Y in both the circuits A and B. As the circuits are very similar in respect to the control and load of the high-voltage transistor the stressor sets of this component will be discussed in one section. All the graphs in this section relate to circuit B, except where mentioned otherwise.

As discussed in Chapter 2 and Appendix A the main function of transistor Y is switching the primary transformer current. See also figures 5.15. and 5.25. This component is analysed in detail because of the considerable differences between the predicted failure rate (using traditional reliability prediction handbooks) and the actual failure rate. As mentioned in chapter 4 the susceptibility limits for this transistor are:

Parameter	Susceptibility limit
Current breakdown	10 A
Crystal temperature	200 °C
Avalanche breakdown	450 V (open base) 1000 V (base-emitter short circuit)
$\dfrac{d\,Ib}{dt}$ (switch on)	$0.5\ 10^{6} A/s$ (slower causes failures)
$\dfrac{d\,Ib/dt}{Ic}$ (switch off)	$K = 5\ 10^{4} s^{-1}$ (slower causes failures) $1.5\ 10^{6} s^{-1}$ (faster causes failures)

Important signals in relation to stressor measurements are in this case the collector current, the base current and the collector-emitter voltage. At the moment of analysis of this component (March 1989 - November 1989) the automatic stressor measurement equipment was able to do only automatic measurements using two measurement channels. (Recent extensions have modified the stressor measurement equipment to measure four channels simultaneously.) Our own experiments have shown that it is possible to use the output of a two channel measurement to obtain an indication of the full stressor set. In those cases where problems exist in relation to the reverse bias second breakdown failure mechanism inadequate base-drive will not only cause an increased susceptibility for this failure mechanism but also a slower switching characteristic. These two effects are closely related. Due to the inadequate discharging of the transistors base (either too fast or too slow) the remaining charge will cause remaining conductivity of (a part of) the transistor. Externally this effect manifests as a decreased slope of the collector current during transistor switch-off. Due to the better indication of failure causes it is suggested a three- channel measurement is used where possible.

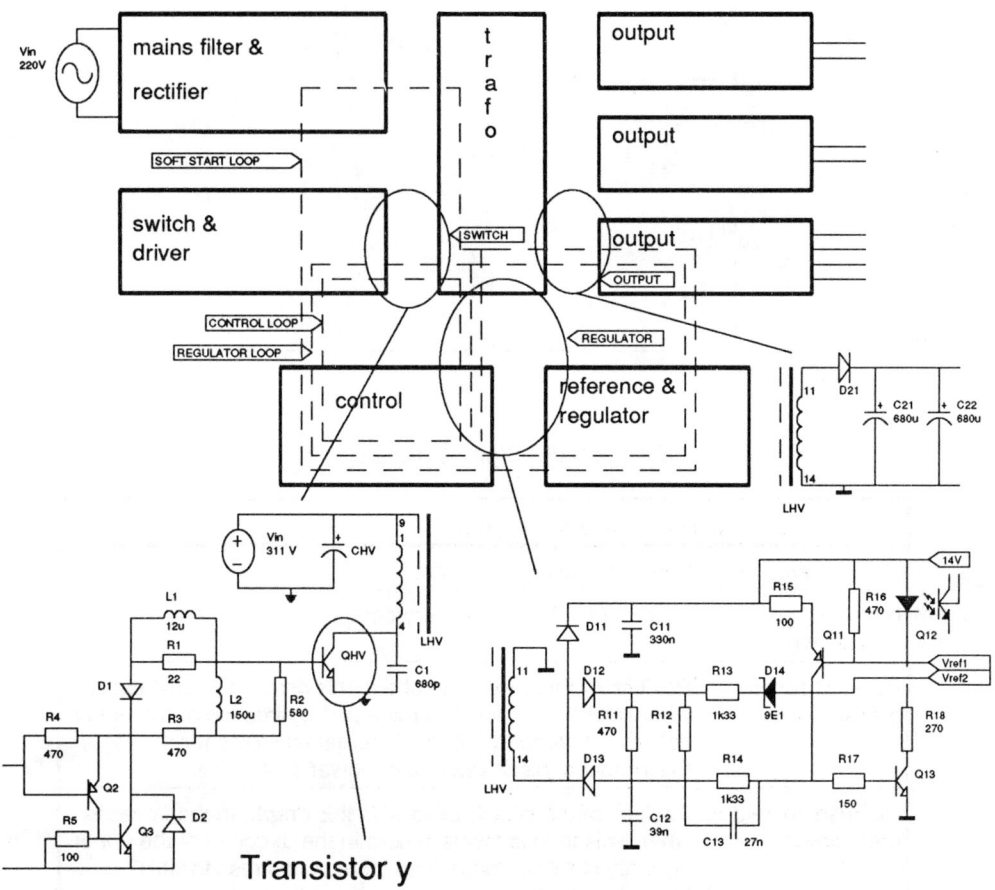

Figure 5.25: Transistor Y in circuit B

5.5.2.1 *Individual stressor set*

As mentioned in the previous section the practical measurements will concentrate on two parameters: the collector current and the collector emitter voltage. Figure 5.26 shows an example of the signals for these stressors.

Very important for the failure mechanisms of this component is especially the relation between collector current and collectoremitter voltage during switch- on and switch-off of the transistor. See figure 5.27. From this figure it is possible to derive an indication for the following failure mechanisms:

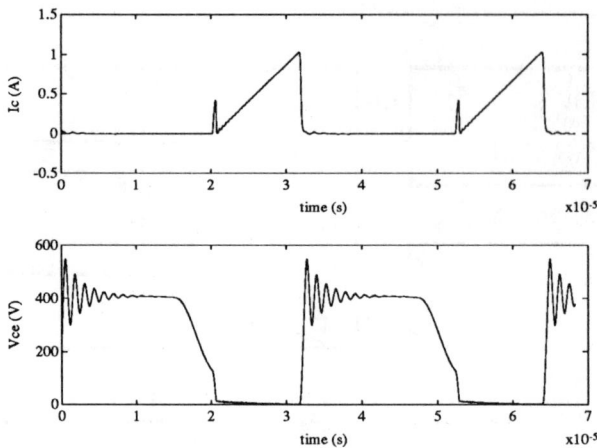

Figure 5.26: Signals transistor Y

Failure mechanism	Relation to V_{ce}-I_c graph
Overcurrent	Maximum collector current
High-voltage breakdown	Maximum collector-emitter voltage
Forward second breakdown	Position of the switch-on lobe in this graph. In those cases where this lobe is located more in the direction of the upper right-hand corner there is an increased stress for this failure mechanism. (Indicates slower switching)
Reverse second breakdown	Position of the switch-off lobe in this graph. In those cases where this lobe is located more in the direction of the upper right-hand corner there is an increased stress for this failure mechanism. (Indicates slower switching)

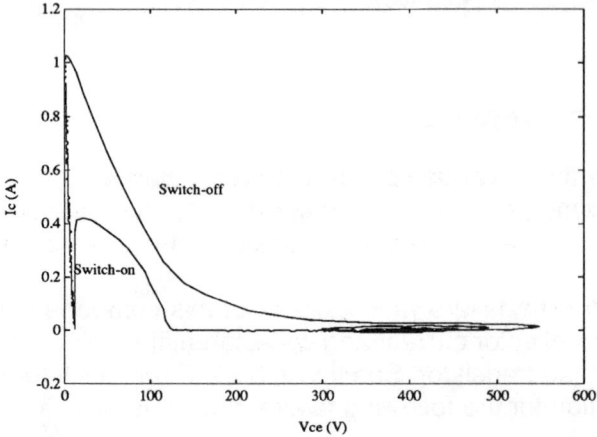

Figure 5.27: V_{ce} plotted against I_c

In a similar way it is possible to present the stressor set for one of the operation modes of the circuit. Figure 5.28 shows the stressor set for the operation mode *play* of the circuit.

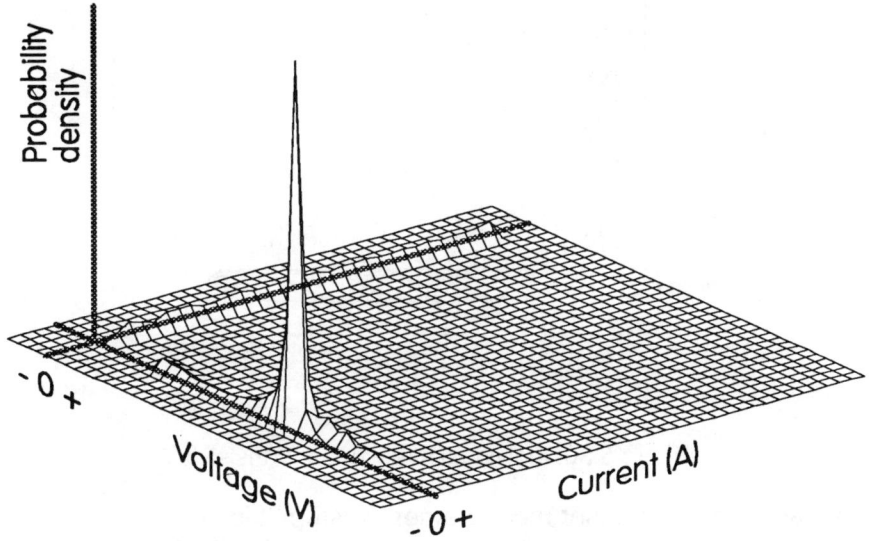

Figure 5.28: Joint stressor probability density plot

The height of the graph indicates the stressor probability density. The main problem with this kind of graph is the fact that especially the (most interesting) peaks (with a low probability of occurrence) of the signal are lost in such a graph. Therefore, as discussed with diode X, an iso- probability density contour is used. See figure 5.29.

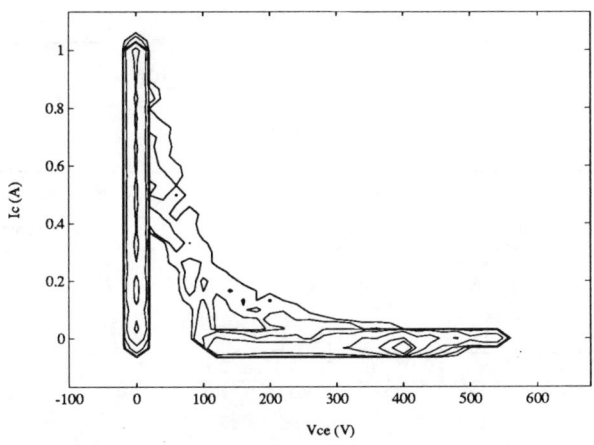

Figure 5.29: Joint stressor probability density contour

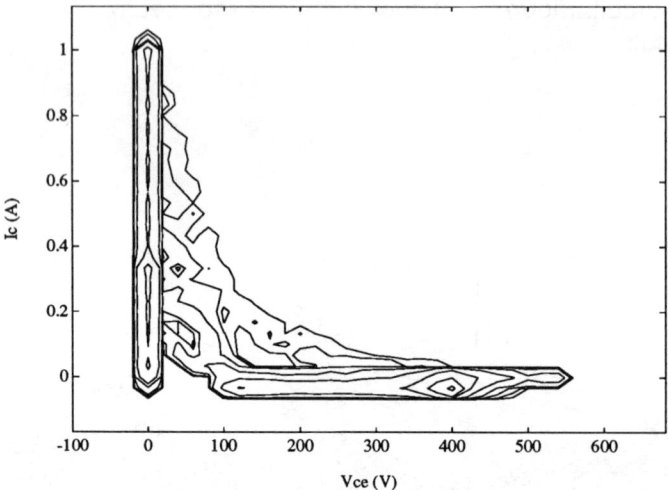

Figure 5.30: Individual stressor set transistor Y

Using an automated stressor measurement system it is possible to derive a full individual stressor set for one single circuit. For a full individual stressor set it is required that not only the stressor sets for all the different states are included but also all the transitions between the various states. The results of the analysis are presented in figure 5.30 which shows the complete individual stressor set for one single circuit.

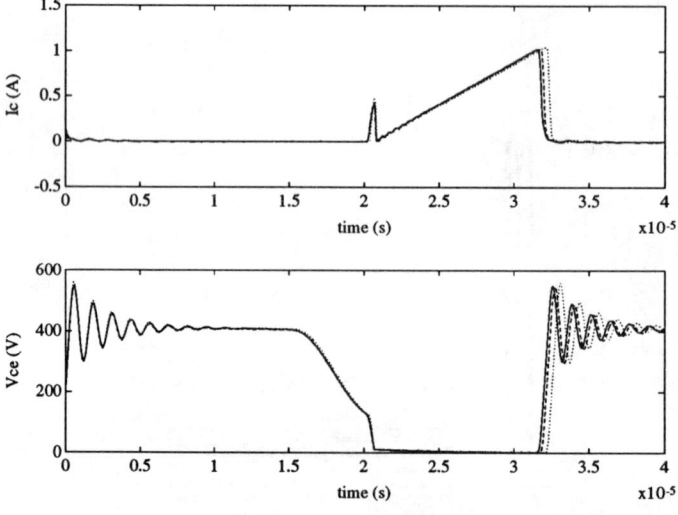

Figure 5.31: Time signal for extreme functional transistors

5.5.2.2 *Tolerance effects of transistor Y*

Due to the feedback technique used circuits arrived for analysis with failed
transistors. In nearly all cases* it is impossible to derive component parameters
for such a failed component. Therefore the effects of functional tolerances of
parameters in transistor Y were analyzed separately. Our own analysis has
shown that one of the most characteristic transistor parameters for this transistor
is the current gain. Many parameters are highly correlated to this current gain.
Use of pre- selected transistors in circuits showed the following differences in
switching behaviour in the time-domain. See figure 5.31.In this diagram it is
difficult to see any differences in behaviour between a transistor with a high gain
(nominal +70%), typical transistors and transistors with a low gain (nominal -
50%). More details can be seen in the V_{ce}/I_c diagram of figure 5.32.

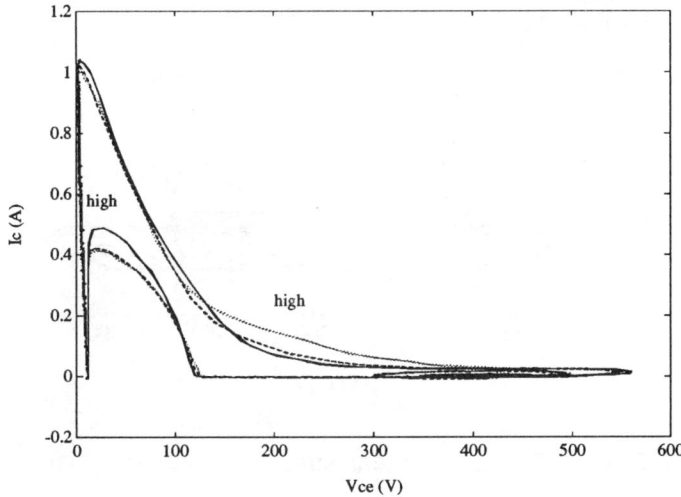

Figure 5.32: Vce/Ic extreme functional transistors

Although it is possible to distinguish clearly the three different transistors in this
graph it is still impossible to decide whether a higher or a lower gain has negative
influence on the stressor diagram. Finally a pseudo** mean stressor diagram

---

*   Using post-mortem analysis it is quite often possible to detect manufacturer failures
        such as poor bondings. In those cases where deviations are not related to such
        manufacturing problems it is very difficult to derive functional parameters of a failed
        component.

**   Pseudo indicates that not the influence of all components was analysed but only the
        influence of tolerance effects of one single component

was constructed, incorporating the influence of three groups of pre-selected transistors and all possible operation modes and transitions between operation modes. See figure 5.34 for a stressor/susceptibility graph and figures 5.33 and 5.36 for the related contour graphs.

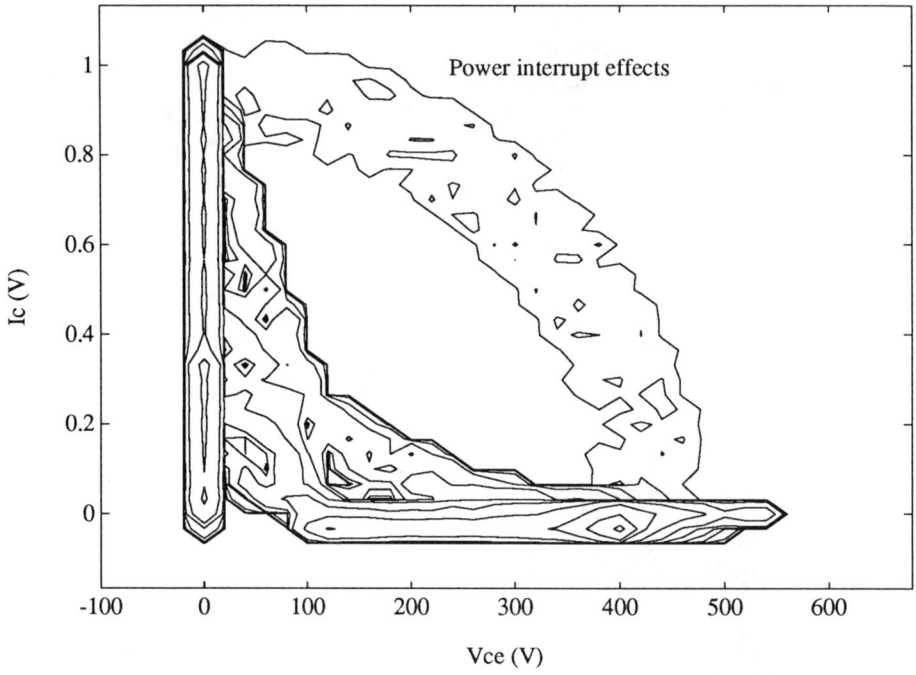

Figure 5.33: Pseudo mean stressor contour diagram

The segment of the curve in the upper-right hand part of this graph is related to the high-gain transistor group during power-interrupts. More detailed analysis of the stressor set during this transition showed that the stressors are beyond the reverse bias second breakdown limit.

This analysis was confirmed by laboratory experiments. Using conditioned power interrupts it was possible to bring many transistors in the high-gain group to failure. Analysis of the burn-marks on the transistors tends to point in the same direction. Burn-marks are especially concentrated in the wide part of the emitter. (See also Chapter 4) Although the extend of the damage on the failed transistors, obtained using feedback, might also include other failure mechanisms, the practical failures confirm failure on reverse second breakdown due to excess stressors. See figures 5.35 and 5.37.

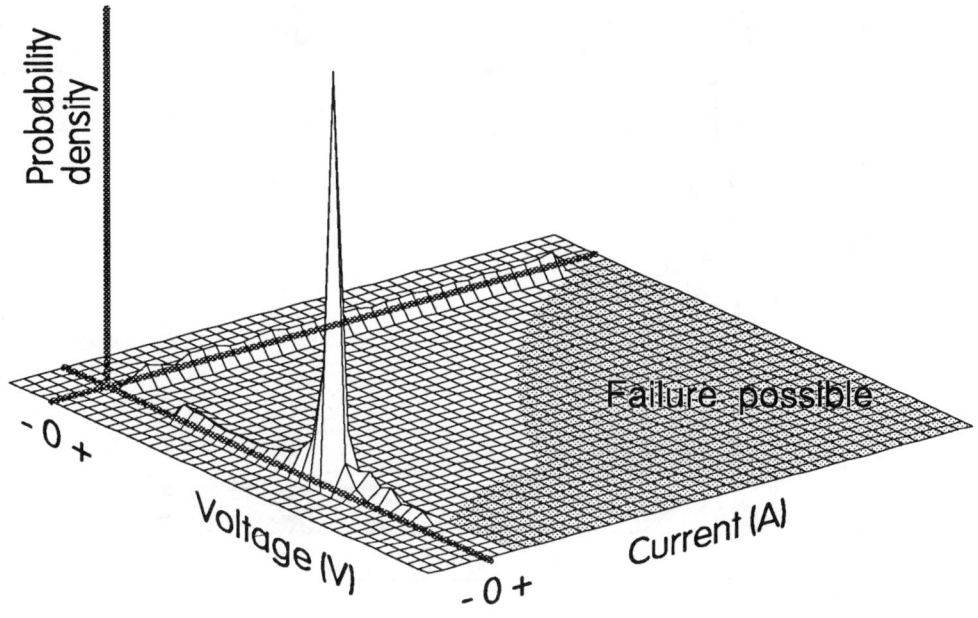

Figure 5.34: Pseudo mean stressor diagram

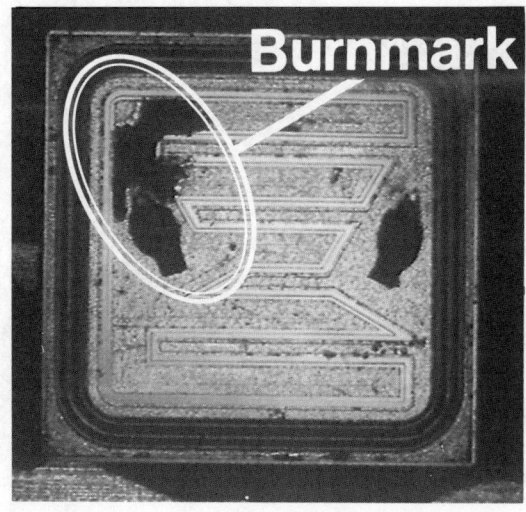

Figure 5.35: Typical burn-mark transistor Y

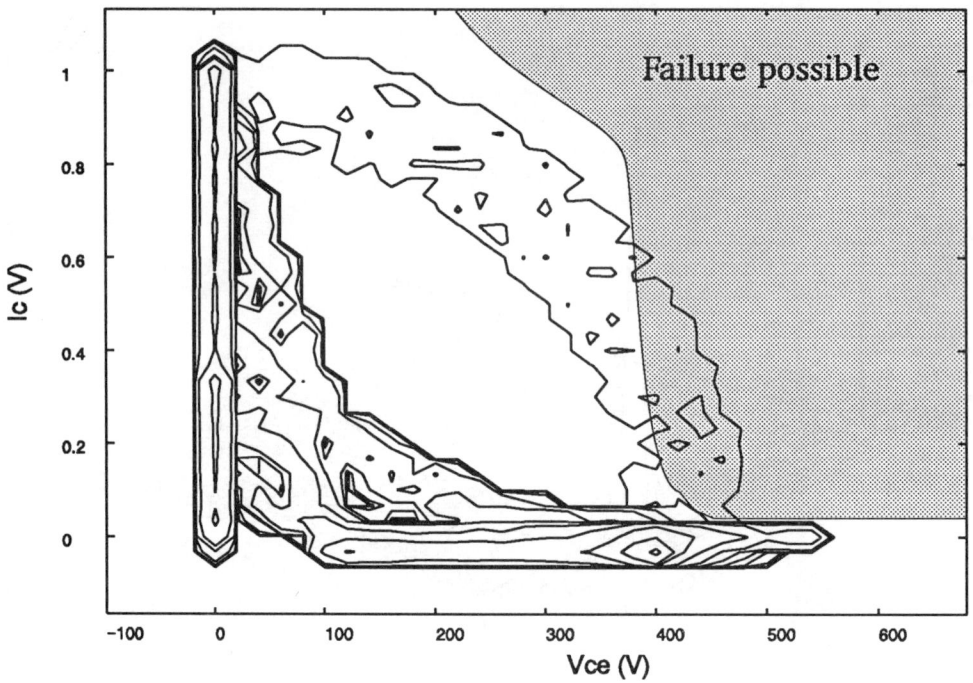

Figure 5.36: Stressor/susceptibility diagram

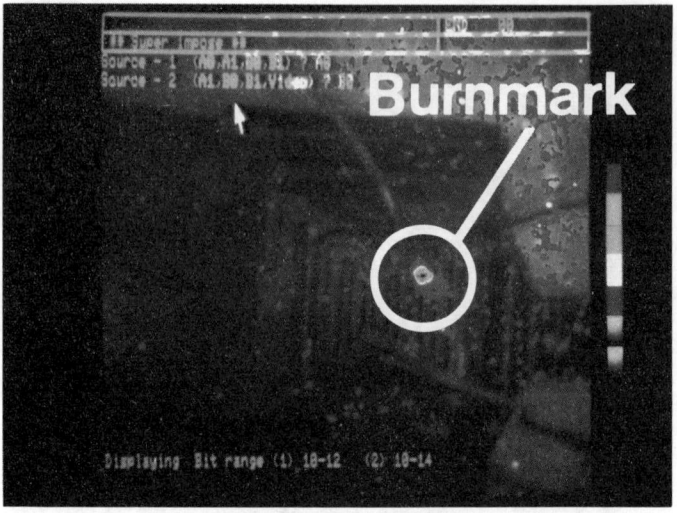

Figure 5.37: Typical burn-mark transistor Y

5.5.2.3 *Tolerance influence of other components*

Using the stressor/susceptibility analysis methods presented in this chapter all circuits obtained using the feedback method were analysed on stressor/susceptibility interaction. For these circuits it was in all cases possible to derive dominant influence factors on stressor sets. The following table summarizes the circuit, the influence factor, the effect and the related failure mechanism.

Circuit	Influence factor	Effect	Related failure mechanism transistor Y
A&B	Transformer saturation current	Excess collector current & voltage	Power breakdown (second breakdown)
A	Optocoupler	Current Transfer Ratio	Instabilities regulator circuit
B	Optocoupler	Optocoupler (in-circuit) delay	Late/not switching off of the transistor

5.5.2.4 *Relation to time-failure probability*

In the previous paragraphs several examples of stressor/susceptibility interaction were discussed:

- — 1: Breakdown voltage diode X

- — 2: Reverse current diode X

- — 3: Reverse bias second breakdown transistor Y
    (related to power interrupts)

- — 4: Transformer saturation current

- — 5: Optocoupler CTR

- — 6: Optocoupler delay

Of the failure mechanisms, related to the above problems, many of the stressor sets are related to two important random factors: the load on the circuit and the environmental temperature. Due to the random nature of the load on a circuit (depending on the use of the system), the environmental temperature and especially, the random nature of power interrupts many of these failures will occur in practice not under pre-defined laboratory conditions but at a random time, depending on the nature of the mentioned random stress factors. This was confirmed by an unintended but very interesting experiment. The power supply of video cassette recorders is full- time operational to guarantee the stand-by functions of the VCR. In the evening of the 23 November 1989 a short mains interruption occurred in the city of Vienna. The next day several field complaints were reported "because the video had suddenly failed".

5.6. **Summary of practical stressor/susceptibility interaction**

The previous chapters have derived susceptibility models for practical components and a method to analyse stressor/susceptibility interaction in circuits. Although the main emphasis of this book is on the integration of reliability analysis in the design process of circuits, the lack of detailed functional (tolerance) models resulted in the deriving of stressor sets from practical circuits. To limit the total number of circuits in the analysis phase the analysis concentrated mainly on circuits showing practical failures. Failing circuits were obtained using a detailed feedback process.

Together the discussed failures occur, for diode X in about 1 % of the circuits, and for transistor Y in about 2% of the population of circuit A and 3% of the population of circuit B which is a vast majority of the actual failures. These failure causes do not indicate a bad nominal design. They do emphasize, however, the need for stressor/susceptibility analysis; if possible in the early phases of the design process. The significant conclusion of this section is the usability of stressor/susceptibility analysis in the deriving of reliability problems. One of the most important conclusions of this chapter is:

> *Using stressor/susceptibility analysis it is possible that for the*   (5.14)
> *discussed circuits many peaks in stressor-sets are related to*
> *functional tolerance effects of component parameters.*

One of the most important problems, described in this chapter, is:

> *Tolerance models of most components are not available.*   (5.15)

An interesting aspect of the actual stressor/susceptibility interaction is the time-dependency of failures due to the random nature of especially the critical stressor-sets. In this chapter the major emphasis is on the analysis of circuits failing in practice. As the next chapter will show, it is possible to use stressor/susceptibility analysis in the early phases of the design process **provided that both component tolerance models and component susceptibility models are available.**

6

Reliability Optimization using Stressor/susceptibility Models

6.1. Introduction

One of the main conclusions of the previous chapters is that stressor/suscepti-bility models are usable for analysis of reliability problems. The remarkable difference between the presented analysis method and traditional reliability analysis methods is that the presented method is able to take into account the differences between circuits within a batch. As Chapter 5 showed the main causes of reliability problems in the analyzed circuits were related to these differences. Reliability optimization therefore requires not only optimization of the nominal circuit but optimization of the entire batch.

It is important to emphasize the difference between batch optimization and optimization of one single nominal circuit. Optimization of one single nominal circuit modifies parameters in such a way, that the nominal design satisfies a number of criteria. Although the name suggests otherwise there are significant improvements of this "optimal circuit" possible (and quite often necessary). Due to tolerance effects a considerable number of circuits may fail. See figure 6.1. This figure shows a certain "safe operating area" of a component as a function of two susceptibility limits. In this example the related stressor function is a function of two parameters, located in the design: A and B. In this case the stressor function is related to the product of A and B; in many practical cases this relation will be far more complex and depending on many more parameters. It is possible to derive from the upper stressor/susceptibility diagram the combi-nations of A and B that will result in a correct circuit and in a failing circuit. This is expressed, as a function of A and B, in the lower part of the figure. Acceptable circuits are indicated with a + and failing circuits are indicated with a -. The nominal stressor/susceptibility interaction remains within susceptibility limits but a considerable amount of circuits has possible failures. Chapter 5 shows several examples of such failures. Therefore batch optimization will require optimization of the circuit in such a way that not only the nominal circuit but also a vast majority of the batch satisfies its demands. This implies that the main demand for reliability optimization of circuits produced in large batches is:

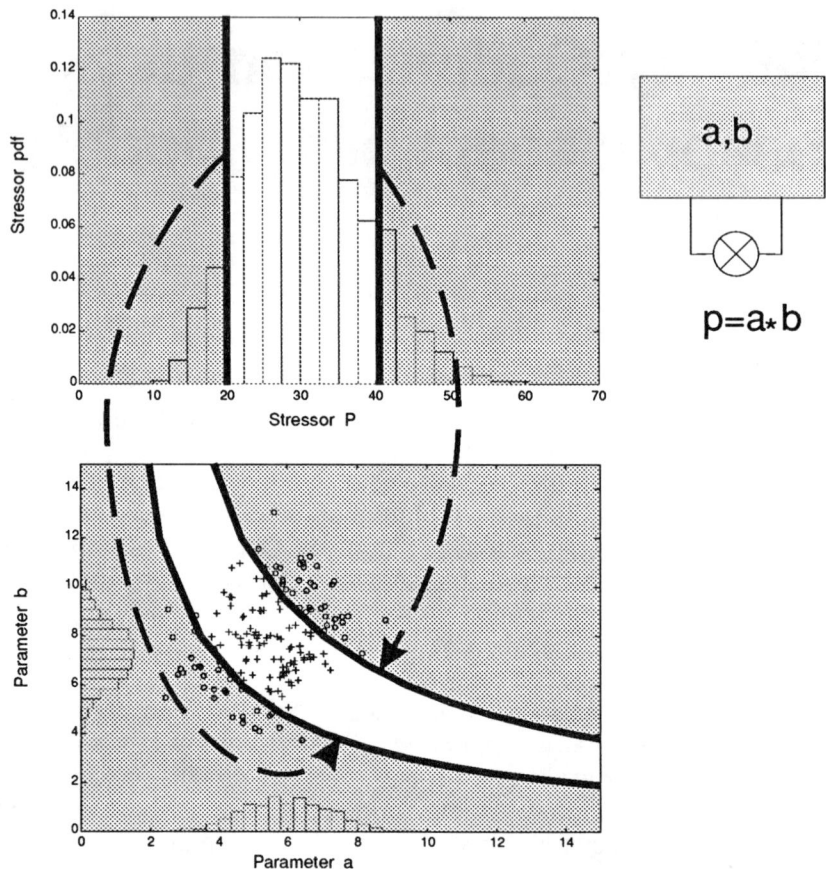

Figure 6.1: Translating stressors to parameters

Reliability optimization requires reassignment of designable (6.1)
parameters in such a way that not only the nominal circuit but
also a vast majority of the batch will perform in accordance with
its reliability specifications.

One of the problems of the method presented is that the majority of the work is related to the development of susceptibility models and/or stressor sets. Although the method provides a relation between designable parameters and reliability it is still difficult to use, as intended, in the early phases of the design process. Reliability optimization requires a relation between the failure probability of a component and designable parameters on circuit level. See figure 6.2.

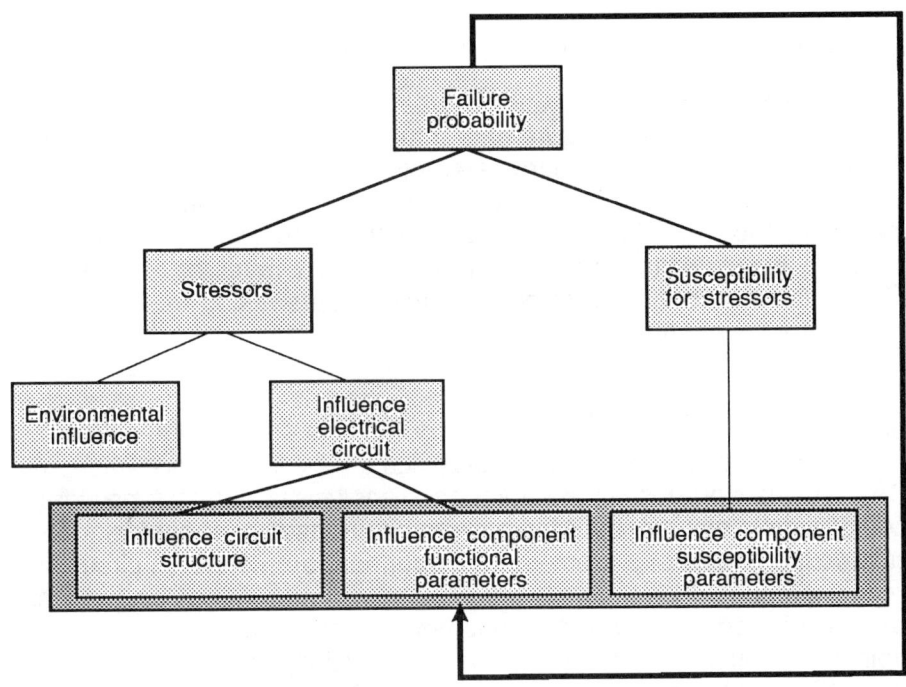

Figure 6.2: Optimization using designable parameters

The major problem for practical optimization remains the unavailability of detailed susceptibility models and detailed tolerance models of many components. This chapter will explain that, under the condition that the required models are available, stressor/susceptibility models are also usable in a (computer aided) reliability optimization process usable in the early phases of the design process. The first sections of this chapter will illustrate the use of Monte Carlo analysis for the simulation and realization of realistic stressor sets. The second part of this chapter will show the use of susceptibility models in the realization of pass-fail diagrams. The third part of this chapter will demonstrate the use of pass- fail diagrams in the parameter optimization process using the centre of gravity method* [Spe88].

---

* The reliability optimization method presented in this book is closely related to the design centering and tolerance technique, published by Prof. R. Spence of Imperial College in London. The reliability optimization methods were developed in close cooperation with Prof. Spence and Mr. Jennings of ISL software as part of this project.

6.2. **Deriving stressor sets from circuit simulation**

One of the conclusions of Chapter 5 was that one of the main reasons for reliability problems (and therefore one of the main aims for reliability optimization) is the effect of functional parameter tolerances on the reliability of a circuit or system. The main problem in analysing tolerance effects of component tolerances on the level of a circuit or system is the often large number of variables that has to be taken into account. Therefore a number of methods will be examined for possibilities to calculate stressor sets taking the effects of part tolerances into account.

6.2.1. *Worst-case analysis*

A well-known method for analysis of tolerance effects is the worst-case analysis (WCA). The main purpose of the worst-case analysis is to determine worst-case values of certain signals. In order to determine these worst-case signals WCA calculates the effects of extremes of component parameters on circuit signals. Extreme component parameters are parameters with values on the limit of a parameter distribution. The problem is how to find especially those parameter combinations causing worst-case signals. One of the possibilities is to use vertex analysis [Spe88a]. Vertex analysis explores all combinations of extreme component values. Given a system with n components each with one parameter described by:

$$p_i = p_{i,0} \pm t_i \qquad i = 1, \dots, n \tag{6.2}$$

In this case circuit analysis is performed for all combinations of parameters $p_{i,0} + t_i$ and $p_{i,0} - t_i$. These combinations form the vertices of an n-dimensional tolerance space. Well-known problem with WCA is the situation that even for small circuits consisting of simple components having only one parameter a complete vertex analysis can become quite tedious. A vertex analysis of a system of n components will require the following number of circuit analysis steps:

$$N_{\text{vertex analysis}} = 2^n \tag{6.3}$$

An additional problem is that the worst-case signal value might be related to a parameter combination which is not connected to one of the vertex points. In other words: worst-case conditions of electrical signals may be caused by other parameter combinations than the mentioned extremes. In case of stressor/susceptibility interaction this would mean that failures may occur for other parameter combinations, not predicted using worst-case analysis techniques. See figure 6.3.

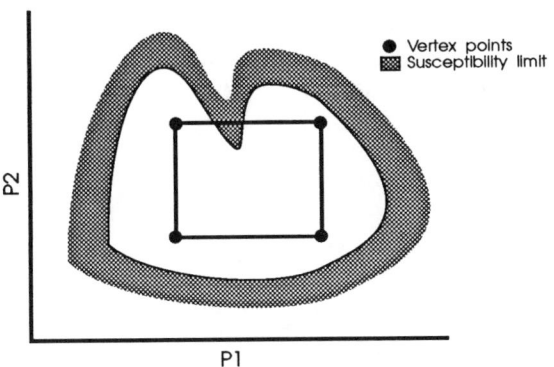

Figure 6.3: Deficiency vertex analysis

6.2.2. *Parameter regionalization*

One of the methods to prevent this omission of parameter combinations is the method of parameter regionalization. Every parameter having tolerances is analysed for a certain number of points, each representing an region within the tolerance space. Each point is assumed to represent the behaviour of its corresponding region. See figure 6.4.

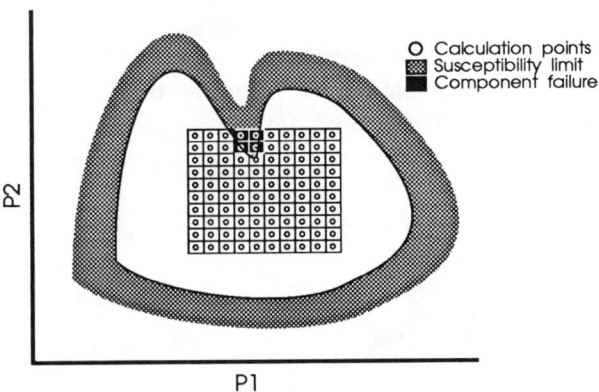

Figure 6.4: Deterministic exploration

Advantage of this method is the smaller probability of omission of failure conditions within the vertices mentioned above, but the disadvantage is the increased number of exploration points. For example: the number of calculation steps for a circuit consisting of n parameters, each split in m regions is:

$$N_{\text{deterministic exploration}} = m^n \qquad (6.4)$$

Even for small numbers of *m* and *n* the total number of simulations required will become uneconomically large. A useful enhancement of this method is related to the fact that of the total number of points analysed using deterministic exploration quite a lot have a low probability of occurrence.

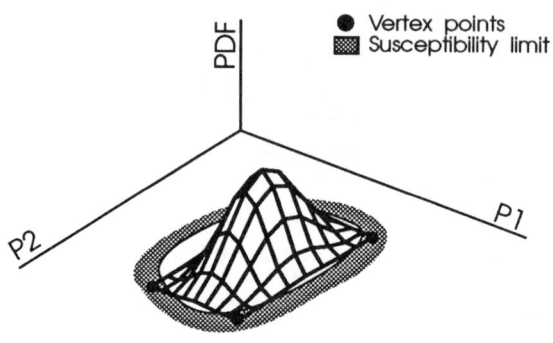

Figure 6.5: Joint pdf with two parameters

Figure 6.5 shows a joint parameter distribution diagram for two parameters. Both parameters are assumed to have a pseudo- Gaussian distribution. The combination of parameter values close to the vertex points has a low probability of occurrence. Parameter combinations with a high joint probability are more likely. To limit the number of analysis steps required by deterministic exploration it is possible to give each deterministic calculation point a weight factor, depending on the joint probability density function of the joint parameter distributions. See figure 6.6.

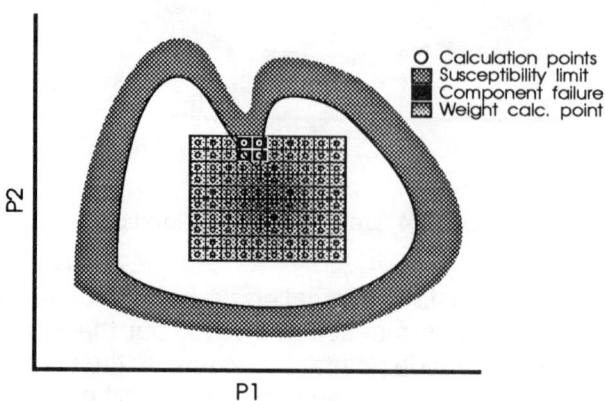

Figure 6.6: Regionalization; exploration using weight

To prevent calculations in hardly realistic calculation points it is possible to analyse the circuit only for calculation points with a weight factor over a certain threshold. This method is called the method of parameter regionalization. [Leu74]. An advantage of this method is the improved coverage factor compared to vertex analysis and the limited number of calculations compared to deterministic exploration. For larger circuits the total number of calculations will be too large for practical applications. (A circuit consisting of 20 parameters, each calculated for an average of 3 calculation points will require already $3^{20} \approx 3.5x10^9$ calculations.)

6.2.3. *Monte Carlo analysis*

Main problem of the previous methods is the problem of dimensionality. In the methods discussed the number of circuit analysis steps is exponentially related to the number of parameters in the circuit. One of the most characteristic common aspects of these methods is the deterministic approach. All methods try to cover the entire tolerance space systematically.

A completely different approach is the method of statistical exploration. Statistical exploration covers the tolerance space by means of the generation of sets of random parameters within this tolerance space. Each set of random parameters represents one circuit. Multiple circuit simulations, each with a new set of random parameters, explore the tolerance space. Statistically the distribution of all random selections of one parameter represent the parameter distribution. Because of its random nature this method is called Monte Carlo analysis (MCA). It is possible to compare Monte Carlo analysis with the production of a series of circuits. See figures 6.7 and 6.8.

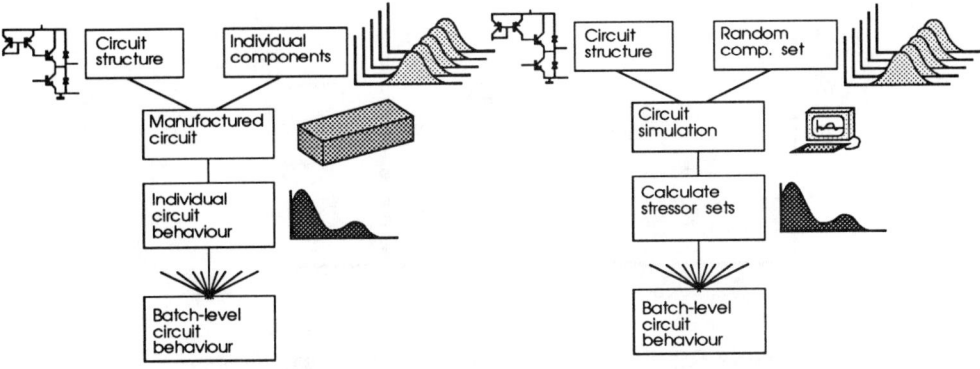

Figure 6.7: Batch manufacturing   Figure 6.8: Monte Carlo simulation

Disadvantage of MCA is the fact that it does not explore the tolerance space in a deterministic way. MCA estimates the effects of parameter tolerances up to a certain degree of confidence, depending on the number of simulations. One enormous advantage is that MCA is not related to the dimension of the tolerance space. In other words, MCA is also suited for problems with many parameters. Although the number of simulations required for MCA is often quite large, this analysis method will be used as a first stage in reliability optimization, especially because the number of parameters in the reliability optimization of circuits (and therefore the dimension of the related tolerance space) is often too large to allow the use of other techniques.

6.2.4. *Pass/fail diagrams*

With MCA it is possible to simulate the behaviour of a large batch of circuits. From this behaviour it is possible to derive stressor sets. The next phase will be the combination of the derived stressor sets with the component susceptibilities in order to decide whether a component will fail or not. With MCA it is also possible to generate susceptibility models for components. As functional parameters of a component can be related to the susceptibility parameters of a component it is necessary to use the same set of random parameters for both the component model and the susceptibility model.

During reliability optimization simplified susceptibility models will be used. As, for the optimization, the most important aspect is to prevent failures, susceptibility will be expressed using the susceptibility limit. See figure 6.9.

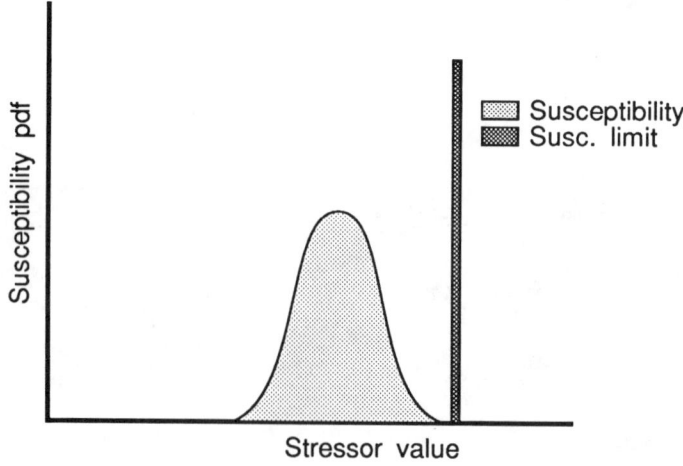

Figure 6.9: Susceptibility approximation using limits

This susceptibility limit assumes catastrophic failure at the moment a stressor exceeds a limit value. This limit value is usually the lower limit of the standard susceptibility function.

Using MCA it is possible by means of circuit simulation to generate a stressor set for a given failure mechanism of a component. To distinguish circuits where failures are possible any circuit in the MCA causing stressors to exceed a susceptibility limit are marked as *fail*. Circuits where no stressors exceed a susceptibility limit are marked as *pass*. Now it becomes interesting to find the relation between parameter values and the fact whether the circuit using that particular parameter value is marked pass or fail. For this purpose the number of passes and of fails is plotted against the initial parameter value. See figure 6.10. In close relation to the statement of Elias for yield prediction [Eli75] it is possible to use the following statement:

> *The more sensitive the reliability of a component is to any* (6.5)
> *particular parameter value the less should be the overlap be-*
> *tween values of that parameter in passing and failing circuits.*

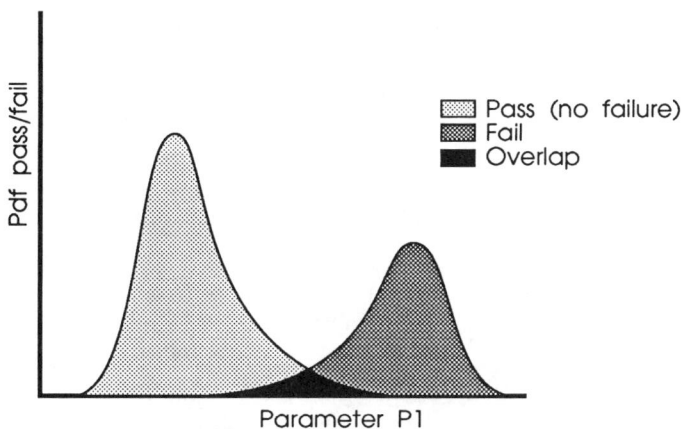

Figure 6.10: Simple pass-fail diagram

In order to calculate this overlap Elias proposed a parameter M_i defined as

$$M_i = \int_{-\infty}^{+\infty} |\; p_{pass}(p_i) - p_{fail}(p_i)\; |\; d\,p_i \qquad (6.6)$$

The value of M can vary between 0 (total overlap) and 2 (no overlap). To estimate the influence factors of a certain failure mechanism it is relatively simple to derive for every parameter in the circuit a pass-fail diagram with corresponding M value. Required are only the output of the Monte Carlo analysis and the related susceptibility limits. See figure 6.11.

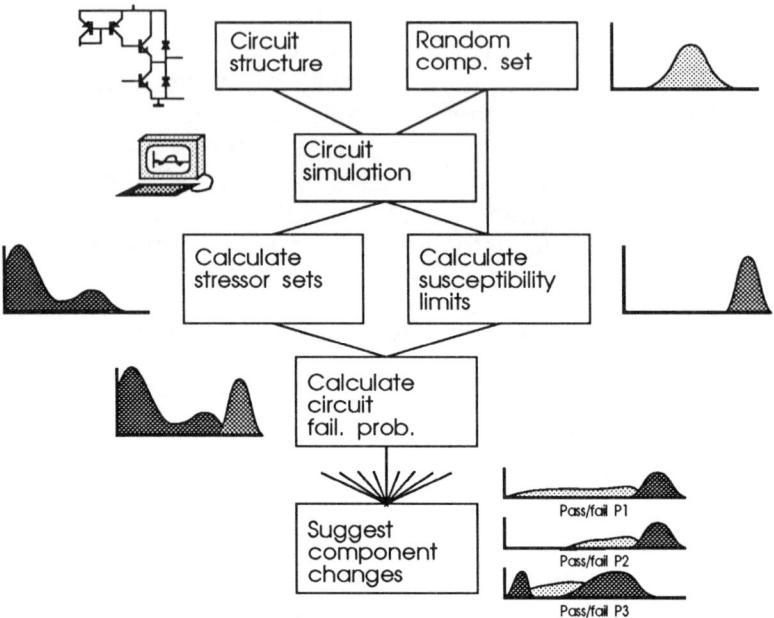

Figure 6.11: Monte Carlo stressor/susc. analysis

This is an important difference with the functional tolerance design methods presented by Spence et al. [Spe88]. The yield, as described by Spence, is determined by fixed specifications. Any circuit outside the specifications is assumed to fail. For reliability optimization the situation is somewhat more complex. Quite often (see also Chapter 4) susceptibility limits are dependent on stressor values. Therefore it is not possible to use fixed limits but it is necessary to decide pass or fail on the stressors but also on the current value of the susceptibility. Another complication is that susceptibility in itself has also random components. Therefore it will be necessary in the generation of the random parameter set to generate not only values for functional parameters but also for the random susceptibility parameters. This chapter will often use figures with fixed susceptibility limits. This is only done as a simplification of reality where many susceptibility limits are not fixed.

Based on the *M* figures of the different pass-fail diagrams it is possible to decide upon a new set of circuit parameters. Options in this respect are the narrowing of tolerances, changing nominal values or changing the circuit structure. For the first two options it is possible to use a semi-automatic process, for the third option effort of the designer will be necessary. As, in all cases, the pass-fail diagrams provide a lot of usable information it is useful to prevent a fully automated optimization. Examples of pass-fail diagrams are presented in figures 6.10, 6.12 and 6.13.

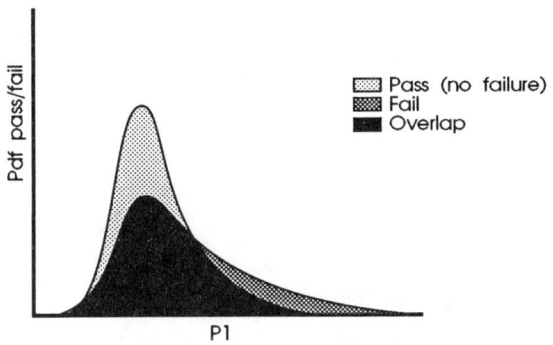

Figure 6.12: Pass-fail diagram; optimization difficult

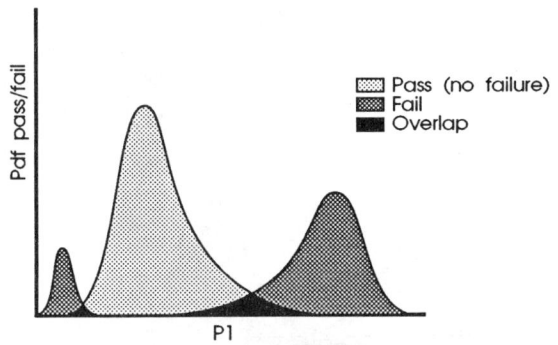

Figure 6.13: Pass-fail diagram; narrow tolerances

Figure 6.10 shows an example of a circuit where a lower value of parameter p1 might show considerable reliability improvements, but in figure 6.12 there is apparently no direct relation between parameter value p1 and failure. In the case of figure 6.13 there is apparently such a relation. In this case narrowing parameter tolerances of parameter p1 seems to be useful. The practical problem will often be: where to start. A circuit of medium complexity will have so many parameters that a straightforward solution is often not available. Therefore the following section will present an optimization algorithm that can be used in the reliability optimization process.

6.3. Reliability optimization using the centre of gravity method

One of the advantages of pass-fail diagrams is that these diagrams are also usable in the process of reliability optimization. It is possible to characterize pass-fail diagrams using two parameters: the centre of passes and the centre of fails. Both parameters are defined as:

$$C_p = \frac{\sum\limits_{i=1}^{np} p_{pass,i}}{np}$$

(6.7)

$$C_f = \frac{\sum\limits_{i=1}^{nf} p_{fail,i}}{nf}$$

(6.8)

where   np = number of passes

                             nf = number of fails

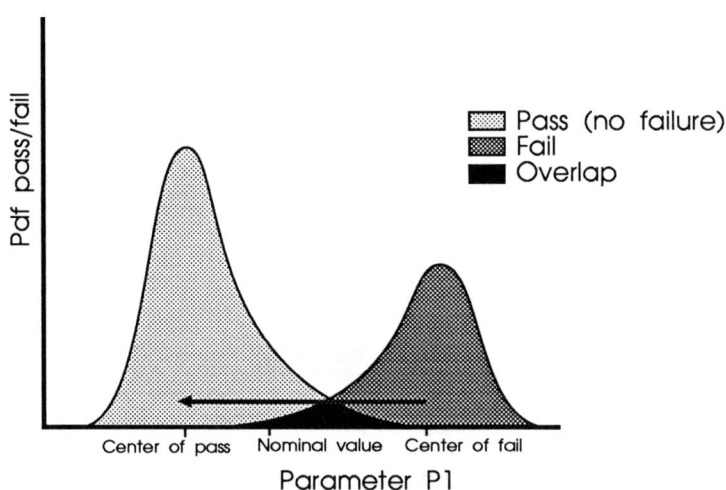

Figure 6.14: Centre of gravity method, one parameter

The so-called centre of gravity method [Spe88b] moves the nominal value along the axis between CP and CF in the direction of CP. See figure 6.14. As the relation between number of passes, number of fails and nominal values is higly nonlinear it will be necessary to use this moving of nominal values in an iteration process in order to obtain (for the reliability) optimal nominal parameter values. Important advantage of this method is its usability for multi-parameter problems. In each iteration step every parameter is moved towards its Centre of Gravity (CoG) until an optimum reliability has been achieved. The only remaining question is: how much should nominal parameter values move per iteration step. The relation between old and new parameters can be expressed as

$$P_{new} = P_{old} + k \cdot (C_p - C_f)$$

(6.9)

In the related field of tolerance design three different policies are used for the calculation of P_{new}:

- $k = 1$

- $P_{new} = C_p$

- $k = 1 - \dfrac{np}{np + nf}$

At this moment there is a tendency to use the third option [Spe88c] . Due to the different nature of reliability optimization, especially regarding the floating nature of the susceptibility limit, experience showed that the second method performs best for reliability optimization.

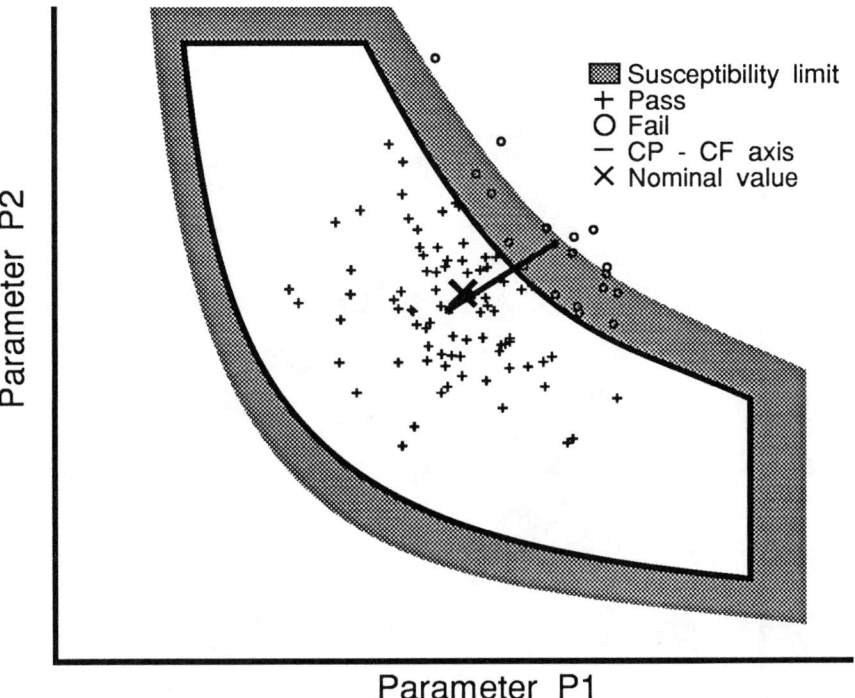

Figure 6.15: Two parameter CoG method; iteration 1

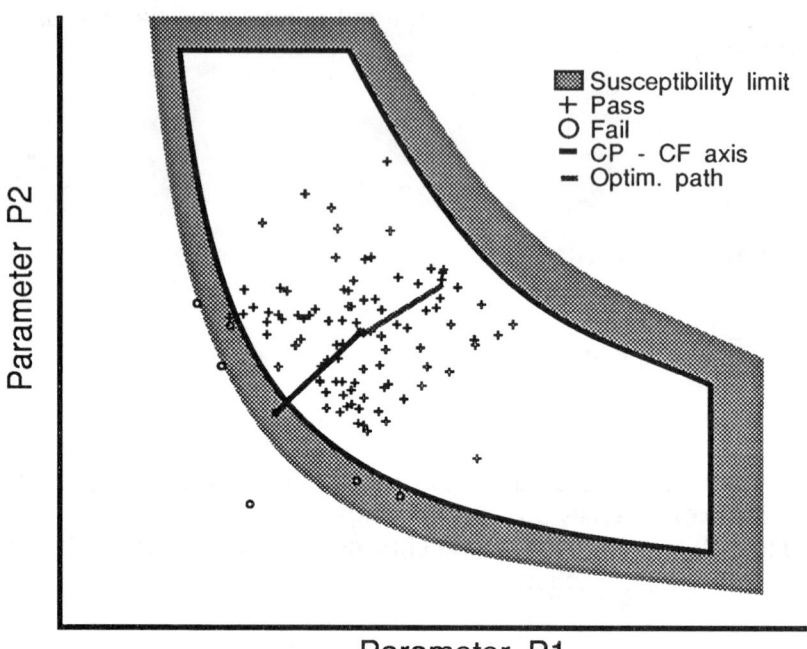

Figure 6.16: Iteration 2

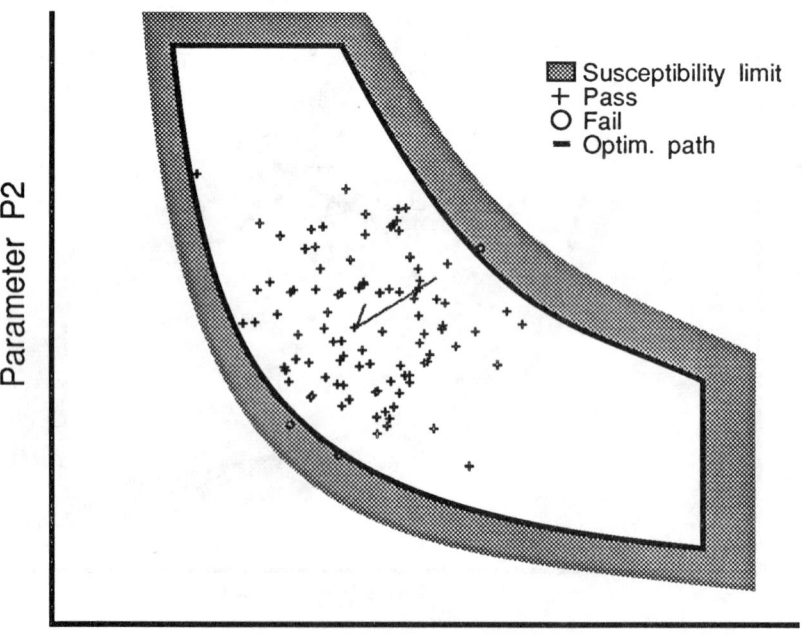

Figure 6.17: Iteration 3

Figures 6.13 to 6.16 show an example of reliability optimization towards two parameters using the CoG technique. Figure 6.1 also shows a WCA solution for this problem. It is obvious that the WCA solution, besides the problems already mentioned in the previous sections, puts greater constraints upon component parameter tolerances compared to the CoG method. The example shows that in a limited number of iterations it is possible to bring a vast majority of the circuits within the susceptibility limits for this failure mechanism. A possible algorithm for the use of the CoG method in reliability optimization is presented in figure 6.18.

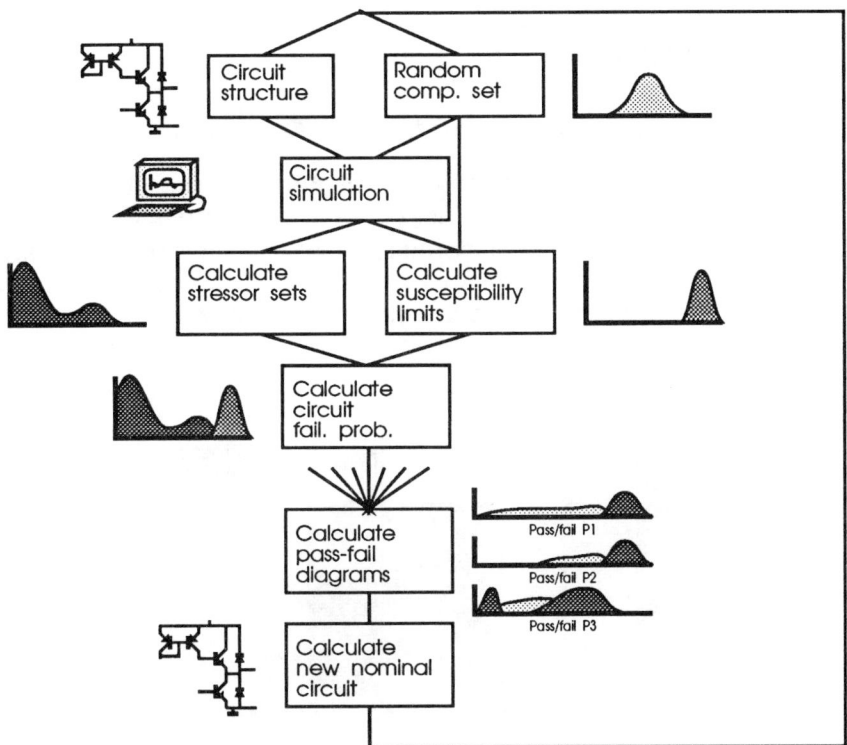

Figure 6.18: Reliability optimization using MCA and CoG

6.3.1. *Long-term stressor and susceptibility models in reliability optimization*

As discussed in Chapters 3 and 4 the susceptibility of a component could be time-dependent. The susceptibility of failure mechanisms like corrosion and electromigration may change with time. Another time-dependent influence is the problem of parameter drift.

These time-dependencies do not fundamentally change the presented reliability optimization method. Only requirements are time-dependent parameter-drift models and time-dependent susceptibility models. Without these models it only possible to do a time-independent reliability optimization. When these models are available it is possible to enhance the presented optimization method. This will result in the possibility to take into account long-term effects (such as degradation, drift and wear-out) in the reliability optimization of products.

Unfortunately practical use of long-term reliability optimization is at this moment not possible due to the unavailability of time-dependent parameter tolerance models. As described in Appendix C of this book many components do not even have a comprehensive tolerance model. See also the section on *models required for reliability optimization.*

6.3.2. *Suggestions for enhancements of the CoG method*

One of the disadvantages of using the CoG method for reliability optimization is the problem that the CoG method is unable to handle circuits where failures do not occur. The optimization is considered completed in those cases where there is no longer overlap between stressors and susceptibility. This implies that the situation presented in figure 6.19 is an "optimal" circuit.

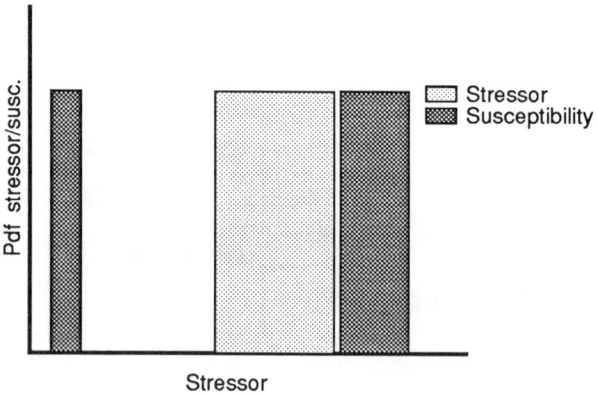

Figure 6.19: "Optimized" circuit close to limits

This optimal circuit, however, may have some considerable disadvantages in practice. The tolerance and susceptibility models used in the optimization process are only limited models. In practice it is possible that the real component parameter distributions will show some, often unexpected, variations compared to the parameter distributions used during the optimization. One of the well-known recent optimization methods, often referred to as Taguchi's method, advocates the use of gradual loss functions in stead of the limit functions used in the previous sections [Kac85] [Bas86] . For reliability optimization this implies introduction of a gradual susceptibility limit as presented in figure 6.20. For

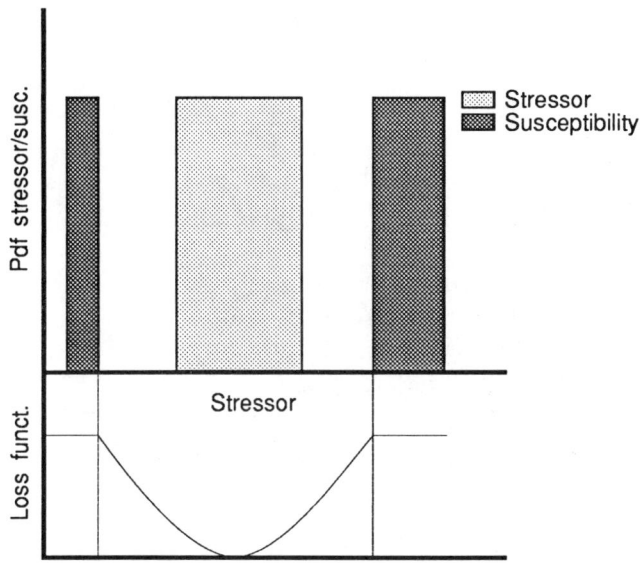

Figure 6.20: Optimized circuit using gradual limits

reliability optimization this could mean the introduction of a new centre of fails. In this case there are two groups of points contributing to the Cf. The first group consists of points beyond the susceptibility limit. These points have a weight factor 1. The second group of points is located within the susceptibility limits, close to the boundary. Depending on the distance to the boundary these point have a weight factor between 0 and 1. (A factor 1 implies a point located on the boundary.)

$$C_f = \frac{\sum_{i=1}^{nf} k_i \cdot p_{\text{fail},i}}{\sum_{i=1}^{nf} k_i} \tag{6.10}$$

Where k_i is a weight factor for stressor point i. $0 \le k_i \le 1$

A further improvement in above optimization method is the introduction of the concept of sensitivity. It might be that a certain stressor function is hardly sensitive to changes in parameter values. See figure 6.21 for an example with one parameter. In this case it is very good possible to realize a design in which the stressor and susceptibility distributions come to a situation of near overlap. In other words: the design is inherently robust.

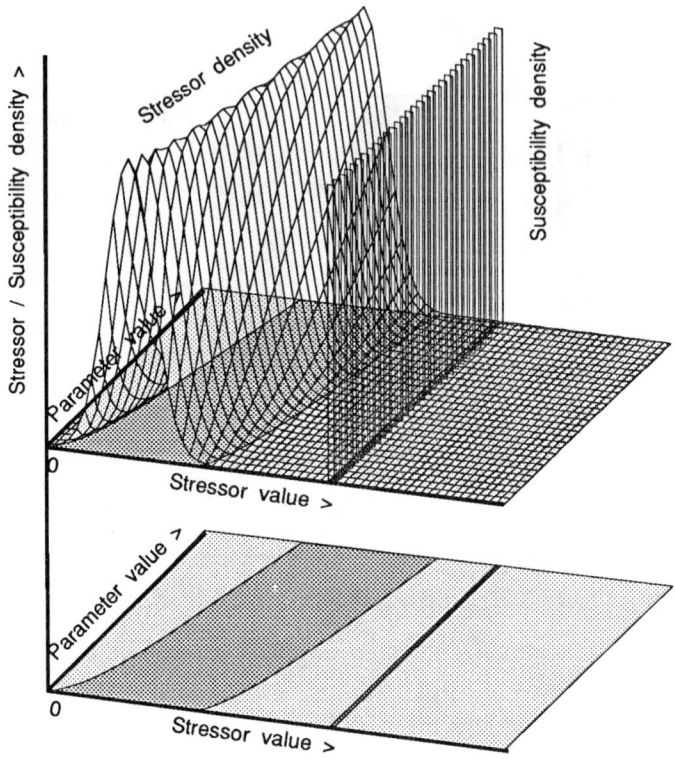

Figure 6.21: Inherent robustness

In other situations the situation might be such that minor changes in parameter values will have a major impact on stressor-susceptibility interaction. In this case the design is highly sensitive. See figure 6.22.

It is clear that the less sensitive situation should be preffered although the margin between stressors and susceptibility can be less in certain situation. In order to design for minimal sensitivity and minimal variability the weight factors will be based on the sensitivity of the stressor-susceptibility function for parameter changes. In many practical cases where a constant susceptibility function is used this can be simplified to the sensitivity of the stressor function for parameter changes.

$$C_f = \frac{\sum\limits_{i=1}^{nf} k(R,\vec{\psi},\vec{p})_i \cdot p_{fail,i}}{\sum\limits_{i=1}^{nf} k(R,\vec{\psi},\vec{p})_i} \tag{6.11}$$

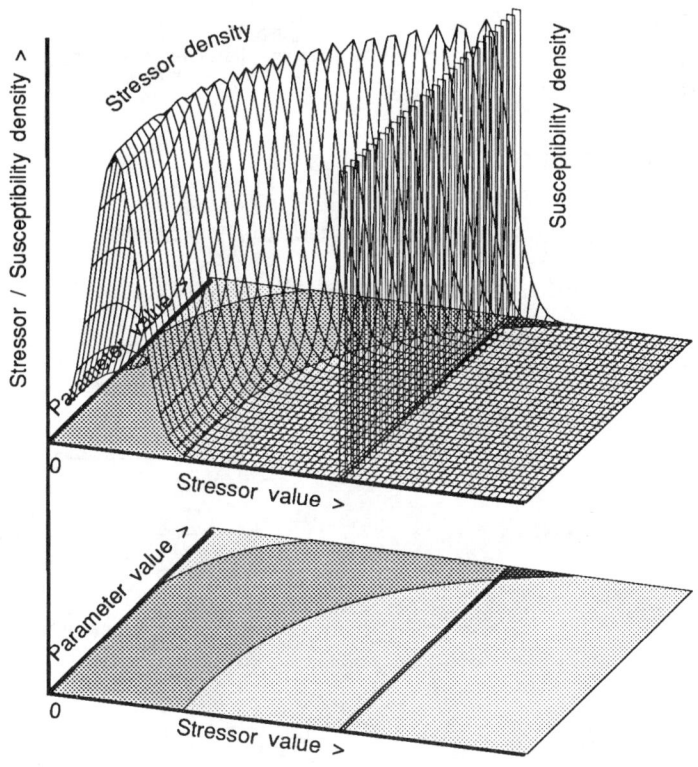

Figure 6.22: Inherent un-robustness

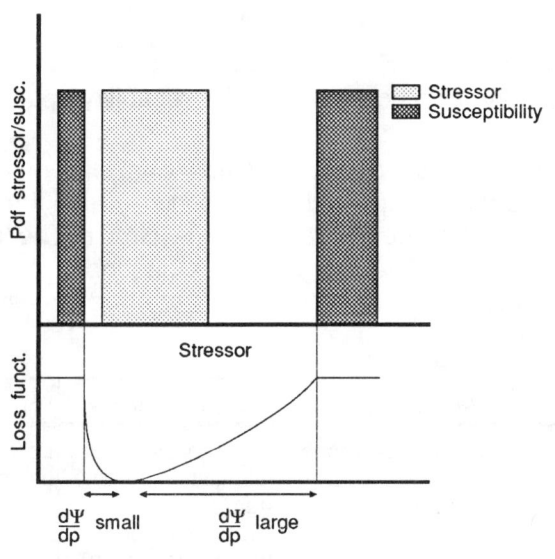

Figure 6.23: Optimized circuit using robustness

Where $k(R,\psi,p)_i$ is a weight factor for stressor point i, depending on the sensitivity of the Reliability (stressor/susceptibility interaction) R, the stressors $\vec{\psi}$ and the (design) parameter vector $\vec{p}$. $0 \leq k_i \leq 1$

See figure 6.23 for a practical example.

6.3.3. *Tolerance models required for reliability optimization*

In the previous sections a method of reliability optimization was described. This

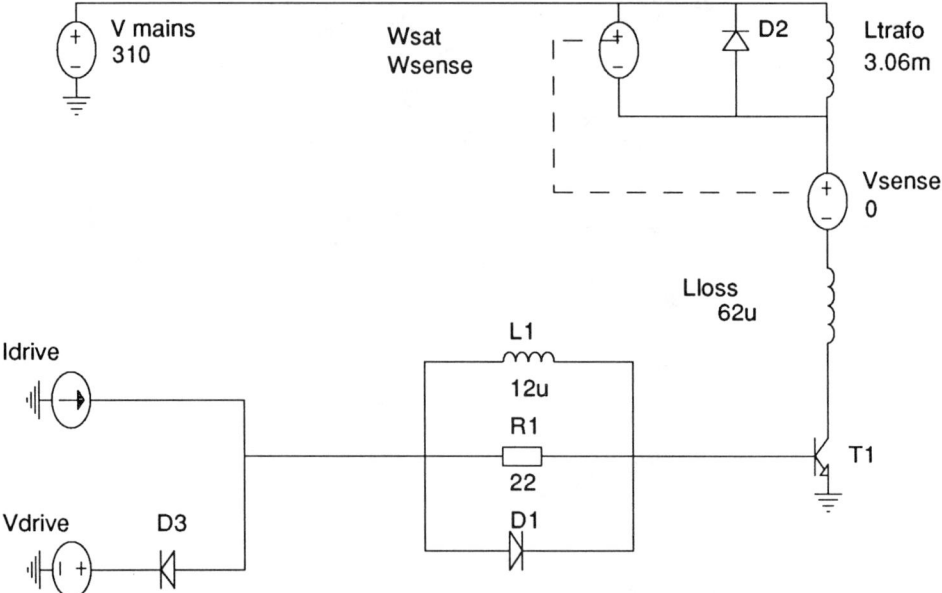

Figure 6.24: Practical example for reliability optimization

method uses a combination of Monte Carlo analysis, used to simulate the behavior and susceptibility of batches of circuits, and the center of gravity method, used to optimize the reliability of the circuit. The method is closely related to the method of tolerance design as described by Spence. Requirements for use of this reliability optimization method are summarized in the following table:

	Nominal circuit	Tolerance effects	Long-term effects
Function (Stressors)	Simulation model Simulation parameters	Tolerances in simulation parameters Correlation between simulation parameters	Drift model simulation parameters

	Nominal circuit	Tolerance effects	Long-term effects
Susceptibility	Susceptibility model Susceptibility parameters	Tolerances in susceptibility parameters Correlation between susc. parameters	Time dependent susceptibility model

See also figure 6.18. As mentioned in Chapter 5 there are some bottlenecks hindering practical use of this method.

— Quite often functional tolerances of component parameters are unavailable (many times only typical values or worst-case limits are presented).

— Correlation models of multi-parameter components are often not available.

— Susceptibility models are hardly ever presented. If susceptibility limits are presented, figures are only presented as partial susceptibility models.

Consequently one of the major parts of this project consisted of the modelling of components. Appendix C gives a detailed overview of tolerance modelling of components, especially of components with multiple parameters.

6.4. **Practical example**

To demonstrate the use of reliability optimization using stressor/susceptibility interaction this section presents a practical example. This example is closely related to the circuits A and B presented in Chapters 2 and 5. See figure 6.24.

The test circuit is a simplified version of the driver circuit of circuit B to demonstrate in detail the effect of tolerances on parameters important for the reliability of transistor T1. As discussed in Chapter 5 important parameters in relation to transistor T1 are:

Parameter	Susceptibility limit
Current breakdown	10 A
Transistor crystal temperature	150 ^oC
Avalanche breakdown	450 V (open base) 1000 V (base-emitter short circuit)
$\frac{d\,I_b}{dt}$ (switch on)	0.5 x 10^6 A/s (slower causes failures)

Parameter	Susceptibility limit
$\dfrac{d\,I_b/dt}{I_c}$ (switch off)	$K = 5 \times 10^4 s^{-1}$ (slower causes failures) $1.5 \times 10^6 s^{-1}$ (faster causes failures)

The simplified circuit is constructed in such a way that similar stressors with similar stressor distributions can be realised. Some unusual components in this circuit are:

— Reverse bias voltage V_{drive}

— Base drive current I_{drive}

— Transformer L_{trafo}

The transformer is modelled in a somewhat unusual way. As the power supply operates according to the fly-back principle, energy is stored in the transformer during the primary phase while the transformer is discharged during the secondary phase. To emulate this discharging effect a diode is placed in parallel with the primary inductance of the transformer. During the fly-back phase this diode dissipates all energy stored in the transformer. The loss inductance is modelled as the independent inductance L loss. Saturation effects are modelled using the current controled switch Wsat. At the moment the transformer current exceeds the saturation current this switch closes, thus emulating the effect of a sudden decrease of induction at the moment of core saturation.

The reverse bias voltage drive Vdrive and the forward base current drive emulate the effects of the transistor drive described in Chapter 5.

6.4.1. *Parameter tolerances*

Main purpose of this example is to illustrate the reliability optimization of a circuit using stress/susceptibility analysis combined with the centre of gravity optimization method. The circuit is optimized for the reliability of the high-voltage switching transistor. For this optimization the following susceptibility model is used (see also Chapter 4):

Main components (with the related functional tolerance model) are:

Component	Tolerance model
Transistor T1	See Appendix C.
Transformer Ltrafo	See Appendix C.
Base coil L1	L tolerance 5 % gaussian

Component	Tolerance model
Base resistor R1	R tolerance 5% gaussian
Base drive current	See chapter 5
Base drive voltage	See chapter 5
Diodes D1, D2, D3	"Ideal diodes"

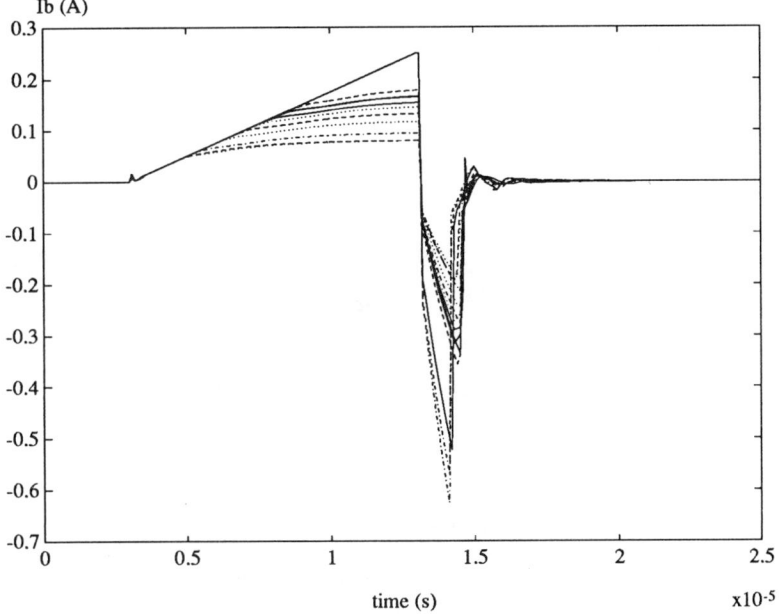

Figure 6.25: Practical tolerance effects on I_{base}

6.4.2. *Optimization using the centre of gravity method*

As discussed in Chapter 5 the main problem in the circuits A and B is reverse bias second breakdown of the high-voltage transistor Y. Therefore this paragraph will concentrate on optimization for this failure mechanism. Figure 6.25 shows the time-diagram of the base current including the tolerance effects.

Figure 6.26 and 6.27 show the (modified) pass-fail diagram for component parameters for this failure mechanism. These figures give also susceptibility limits for this failure mechanism.

It is clear that especially the reverse drive voltage is usable for optimization. Use of the CoG method on all the parameters described above gives the result that optimum reliability can be achieved using the following modifications [Bro89c]:

— Make upper bound of V_{drive} more negative (< -4 V)

— Make lower bound of I_{sat} transformer higher (> 1.5 A)

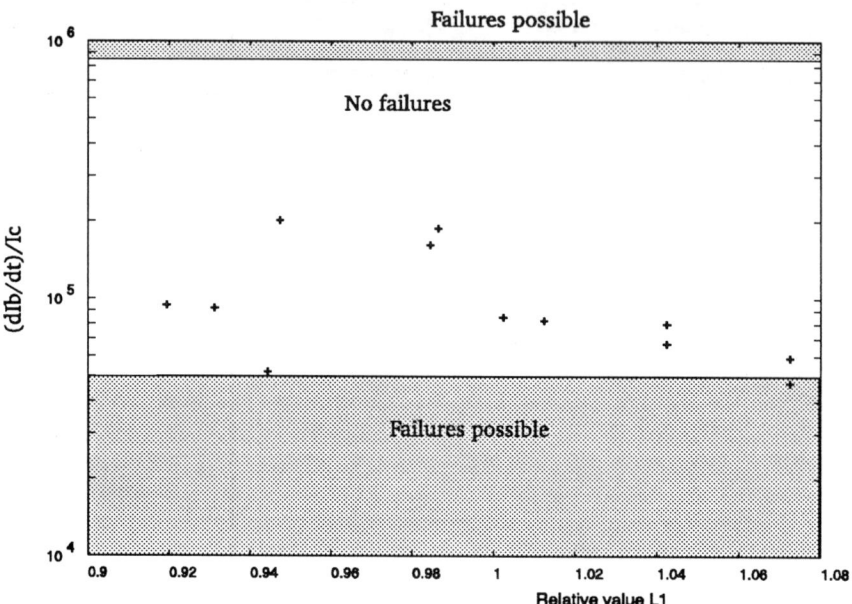

Figure 6.26: Relation L1 / susceptibility limit

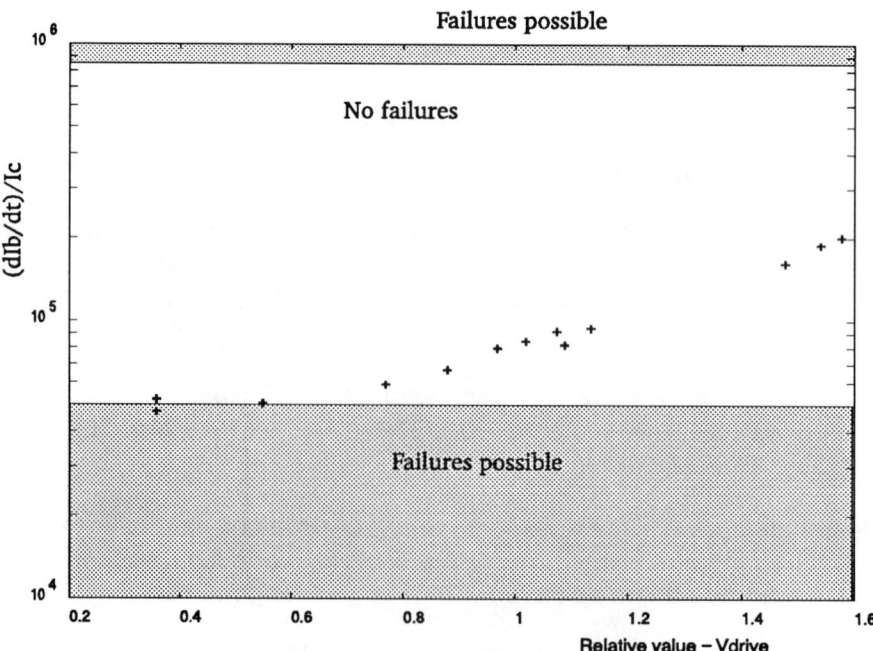

Figure 6.27: Relation Vdrive / susceptibility limit

— Check especially the transistor parameters R_b and C_{jc}. The high values of these parameters show more failures.

— Limit current I_{drive} (harmful, especially in combination with a high C_{jc} transistor and a low V_{drive})

Although this is a very simple example it shows that it is possible to use the CoG in the reliability optimization process. The optimizations mentioned above do confirm the practical analysis discussed in Chapter 5. Advantage of this optimization is the fact that it is based entirely on computer simulation and not on the considerable number of circuits failing in practical situations.

7
Conclusions

The initial aim of this book was the adaptation of existing- (or the development of new)- methods and models for reliability analysis usable as part of the design process of electronic circuits and systems. To demonstrate the usability of the methods presented they were verified on a number of practical circuits.

7.1. Impossibility of using existing reliability prediction methods

The analysis of existing reliability prediction methods showed that the vast majority of the methods in use is based on the principle of part failure rate prediction. Although the predicted failure rate value depends very much on the model used (the predicted failure rates for one component may show variations of over a factor 100, depending on the used prediction model) there is one common aspect in most prediction methods. All prediction methods assume failure rate figures to be thermal dependent in accordance with the Arrhenius law. The activation energy, one of the most important terms in this law, is very different for the various models. Most part failure rate prediction models have one or more parameters related to the internal construction of the component. Correction factors are also presented to take into account the effects of application of components under different environmental conditions. The part failure rate prediction assumes no mutual influence between parts. Only the influence factor related to the electrical environment is the stress factor. This factor is often related to the thermal effects of power dissipation in the component. Only in some cases are correction factors introduced for the application in which a component is used. None of the available part reliability models assumes differences between individual parts within a batch.

There are several problems preventing the use of the currently available failure rate prediction models in the reliability design optimization. First of all the current models lack parameters related to *designable* parameters. The majority of the designable parameters in the design of electronic circuits is in one way or another related to electrical signals. The only common factor in the current generation failure rate prediction models is thermal stress caused by environmental temperature and power dissipation, in spite of the fact that electronic components are known to be susceptible to more failure mechanisms. In those cases where failure rate prediction models present other influence factors these factors are given in the form of correction factors. A transistor with a switching function is, for example, assumed to be more reliable than a similar transistor used in a linear application. As part failure rate prediction models are based on wide-range feedback of many circuits operating under a vast number of different conditions. This difference between linear and switching devices will undoubtedly be true on a statistical basis but it will hardly be of any use in the design of one single circuit. Another problem with the existing failure rate prediction is the enormous difference in predicted values. It is difficult to base an optimization method on a technique where the predicted values for one component are totally different for different models.

7.2. New method: stressor/susceptibility interaction

A new reliability analysis technique was therefore developed. This new technique is based on the susceptibility of failure mechanisms in individual components for external influence factors called stressors. Important difference between the presented technique and traditional reliability prediction methods is that this method is based on the individual susceptibility of components for combinations of stressors (also called stressor sets) occurring in individual circuits. Although the new method shows some similarities with the stress-strength method presented by Jensen [Jen89], an important difference is the use of multidimensional individual stressor sets. Reliability optimization does not use analysis of one failure mechanism with one stressor but requires for one component the analysis of all failure mechanisms with all stressor sets. Only then is it possible to predict component failures.

Another difference with traditional reliability prediction methods is the use of the factor time. Many researchers have found that the constant failure rate assumption is only valid for a certain part of the lifetime of a system. Usually shortly after production the failure rate is considerably higher than the assumed constant failure rate period during the "useful life" period. At the end of the lifetime the failure rate increases, due to what is called "wear-out" failures. In terms of stressor/susceptibility models it is possible to explain this bathtub curve using weak sub-populations failing shortly after productions, a constant and identical population (under constant stressors) during useful life while degradation causes wear-out failures. An important advantage of the stressor/susceptibility method is the possibility to explain deviations from the bathtub curve such as observed by, for example, Jensen [Jen 89].

7.3. **Practical use of stressor /susceptibility models**

The new method was tested on two practical circuits*. Using stressor/suscepti-bility interaction it was possible to explain a majority of the practical failures occurring in this circuit. It was possible to explain both failures shortly after the production of a system and long term failures. It is remarkable that the long-term failures were especially related to the long-term characteristics of the stressor distributions and not of the susceptibility distributions. (This may be different with other circuits.) It is interesting that long-term stressor characteristics are not taken into account in one of the traditional reliability methods. The method presented in this book appears to be useful especially for those circuits where stressor/susceptibility interactions can be expected. This is especially the ca-tegory of circuit where high power, high voltage and/or high current effects are important.

7.4. **Development of susceptibility models**

One of the most important steps in stressor/susceptibility analysis is the devel-opment of susceptibility models. During the development of susceptibility mod-els it turned out that many models concerning failure mechanisms are available. Most models are usable as basis for susceptibility models but the practical application of these models is often quite difficult. Therefore in most cases modelling was based on, or verified by, practical measurements. Although limitations in the measurement equipment prevented the development of com-plete statistical susceptibility models it turned out that it was possible to derive "safe" susceptibility models. The term safe in this case indicates susceptibility models in the form of worst-case susceptibility limits. Although the development of complete statistical susceptibility models was not yet possible, the develop-ment of these models deserves attention for further research. At this moment manufacturer's databooks tend to present only comparatively simple operation guidelines without distribution and often even without distribution limits. The introduction of detailed susceptibility models in component manufacturers data-books appears to be a useful enhancement of these books and gives additional insight to the user of these databooks. An additional advantage of a manufac-turer providing susceptibility models, is the possibility to introduce the models in a computer aided design system, thus improving the analysis speed.

---

*   Analysis of a third circuit was not completed due to the unavailability of sufficient test
     samples for more detailed statistical analysis.

7.5. **Stressor sets**

Generally speaking there are two methods to find stressor sets for a component: measuring stressors and obtaining stressor sets from the results of computer simulation. In this book measurements were used most of the time. This especially due to the unavailability of usable computer simulation models. Although many component models are available for programs such as Spice and Philpac there is, at this moment, no generally accepted tolerance model for the more complex multi-parameter devices (such as diodes, transistors, etc.). Although many computer programs have some rudimentary possibilities to introduce parameter tolerances, these tolerances are in practice hardly known and correlations between parameters are not known at all. Therefore a considerable effort was put in the development of more comprehensive tolerance models for especially multi-parameter components. The reason for the emphasis on tolerance models is that practical analysis showed that an important part of the peaks in practical stressor distributions was closely related to extremes in the functional tolerance distribution. Analysis of circuits A and B proved that a majority of the reliability problems was related to extremes in the stressor function. For this type of circuits the use of stressor/susceptibility analysis in the early design phases would have been quite useful.

7.6. **Reliability optimization**

The stressor/susceptibility analysis method described in this book has some similarities with the functional tolerance design techniques thus it appeared useful to adapt functional tolerance optimization techniques for the purpose of reliability optimization. An important difference in this respect is the fact that functional optimization uses fixed specification limits while stressor/susceptibility analysis uses, due to the nature of susceptibility, a variable susceptibility, depending on the individual characteristics of a (randomly selected) component. An important limitation in this respect is the current unavailability of more complex functional and susceptibility models for many components. Practical experiments with reliability optimization using the centre of gravity optimization method have shown that it is indeed possible to use reliability optimization as a part of the functional circuit synthesis. Using the realized reliability optimization software, developed by ISL software in London, U.K. it is possible to prevent a majority of the failures occurring in the analyzed practical situations already during system synthesis. The most important bottleneck at this moment is the unavailability of adequate tolerance- and susceptibility- models for many components.

7.7. **Practical use of stressor/susceptibility analysis in industry**

As mentioned in the earlier sections one of the major problems of the use of computer simulation in actual reliability optimization is the current lack of actual accurate simulation models. Therefore, as mentioned, many practical stressor sets (and the related susceptibility models) where derived from practical measurements. Although it is possible to use this "measurement based" form of analysis and optimization in actual projects the results of the measurements are not only used for the optimization of single projects but also for the development and enhancement of a library of both higly detailed functional simulation models and susceptibility simulation models. The use of this combined measurement - simulation approach leads to a knowledge base of past reliability experience. In this knowledge base the knowledge is stored implicitely in both categories of simulation models. Although a first project on a new category of circuits will result in a slow and tedious analysis (both the practical measurements and the related model development should be carried out) latter projects can be carried out much faster and, through the available models, prevent mistakes made in a first circuit being reintroduced in later generations of the circuit. See figure 7.1.

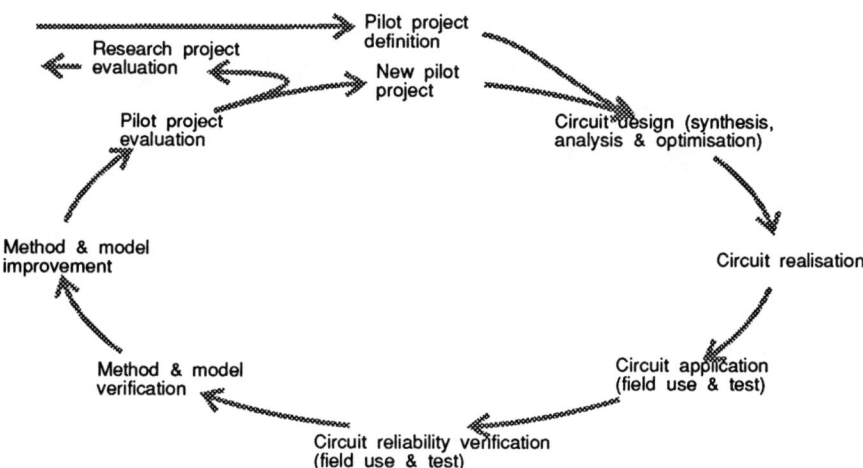

Figure 7.1: Development of models using pilot projects.

7.8. Recommendations for further research

The presented stressor/susceptibility analysis method is at this moment especially useful for circuits with relative high power dissipation, high voltages and/or
high currents. The requirement for practical use of this method is the availability
of sufficient detailed tolerance- and susceptibility- models. Publishing tolerance
and susceptibility models would be a useful enhancement in component databooks. The circuit, mentioned in this book as circuit C, was not analyzed in further
detail due to the unavailability of sufficient samples for detailed analysis. Further
analysis of this type of integrated circuits appears to be useful, especially
regarding the vast differences in predicted standard failure rate figures. These
differences, combined with the random nature of the load of this IC, indicate that
also for this category of circuits there might be an interesting interaction between
component susceptibility and stressors. Finally the practical susceptibility models were considered time-independent. An interesting area of research provides
the analysis of influence of time-dependent susceptibility limits, such as caused
by, for example, electromigration, on the actual time-dependent failure behaviour.

Appendix A
Short Explanation of the
Test Circuits Used

A.1. Introduction

To develop and test the methodology for the optimization of reliability of electronic circuits and systems, practical experiments were carried out during the time-interval October 1987 - March 1990. During this period three different circuits were analyzed. All circuits were used in Philips video cassette recorders. These three circuits were selected for a number of reasons.

— All are circuits where traditionally reliability problems can be expected.

— Most (key-) components within these circuits were produced within the Philips concern so that it was possible to have access to the required detailed information.

— The circuits were produced in sufficient quantities (~1500000 systems / year) to make adequate feedback possible.

Although the circuits used come from a limited application range, in most cases the results obtained will also be applicable for other classes of circuits. It is important to know that the circuit description presented in this appendix is not intended for any other purpose than to clarify functional backgrounds, necessary to explain reliability aspects discussed in this book. More detailed analysis on the level of failure causes is presented in [Bro89a] and [Bro89b]. The following sections discuss functional aspects of the test circuits used. These sections also present reliability figures for the components used in these circuits using traditional reliability analysis/prediction methods.

A.2. Practical circuits

To test the usability of methods, commonly found in standard reliability prediction handbooks, in the process of reliability optimization the following practical circuits were used:

— Circuit A: self oscillating power supply.

— Circuit B: self oscillating power supply.

— Circuit C: motor drive circuit.

All circuits were used in Philips video cassette recorders. The practical failure rate figures, used for verification purposes, are based upon field data obtained during the years 1987 to 1989. More details concerning the actual field failures are presented in Chapter 5. The examples focus on certain components as these components are used for a more detailed analysis in the other chapters of this book. The circuits were selected because company experience showed that reliability problems were expected especially in power circuits. The examples will show only the most important parts of the circuits. The actual calculations and the verification measurements were performed on the entire circuit.

A.3. Reliability predictions for circuits used in consumer electronics

Before it is possible to estimate the reliability of a certain component it is important to know the values for the parameters required by the various reliability prediction models. As can be seen in Chapter 2 of this book, it is possible to distinguish the following parameter groups:

— Parameters related to the standard environment class

— Parameters related to the used component screening techniques

— Parameters related to the device structure

— Parameters related to the application class

— Parameters related to the effective device temperature

In this table the parameter groups are (roughly) increasingly determined by the application.

As the circuits in which the components are used are intended for use in consumer electronics, it is easy to derive parameter values for the first two parameter groups. Generally speaking, systems such as video cassette recorders are intended for use in a protected (living-room) environment. This implies that the environmental conditions are less than ideal with some environmental stress and limited maintenance. This can be derived from the conditions for use of the equipment.

For all circuits the following environmental requirements are specified:

— Ambient temperature: +15 - +35 ^{o}C

— Relative humidity: 30 - 80 %

— Position: fixed, horizontal angle 15°.

This implies that large climatological changes are not expected. It is also unlikely that parameters refering to portable equipment should be applied. Another aspect of high-volume consumer electronics is that, under normal circumstances, screening techniques resembling the military screening classes are not applicable. Together, this leads to the following assumption for environmental and quality screening parameters:

Parameter	Meaning	Selected MIL class	Selected BT class
π_e	Environmental acceleration factor	Ground fixed	Ground fixed
π_q	Quality acceleration factor	component screening classification	lower (minimum) screening class

Further details of the circuits used and components are discussed as part of the discussion of the corresponding circuit.

A.4. Circuit A

Figure A.1 shows the functional structure of circuit A. Circuit A is a self oscillating power supply intended for use in video cassette recorders (VCRs).

The power supply is designed to be able to deliver about 35 W output power. It is intended to fulfil the following functional demands:

Parameter	Functional demands
Input voltage	170 to 265 V rms, 45 to 65 Hz

Parameter	Functional demands	
Output voltage	5.2V	280 - 800 mA
	5.2V	800mA
	14.5V	120 - 500mA
	14.5V	40 - 850mA 1600 mA (<5sec.)
	33.5V	0 - 4 mA
	-29.0V	15 - 120 mA
	3.8V ~	100 - 140 mA~
Operating frequency	20 - 120 kHz	
Efficiency	78 %	

This chapter will not discuss the very details of this circuit. The functional outline of the circuit will be presented in the following section. Only where functional aspects are important for the reliability of (parts of) the circuits will they be discussed in detail.

A.4.1. *Functional structure circuit A*

Figure A.1 shows the rough structure of the supply. The lines in figure A.1 indicate the connections between the various modules. Most important modules are given in the table below.

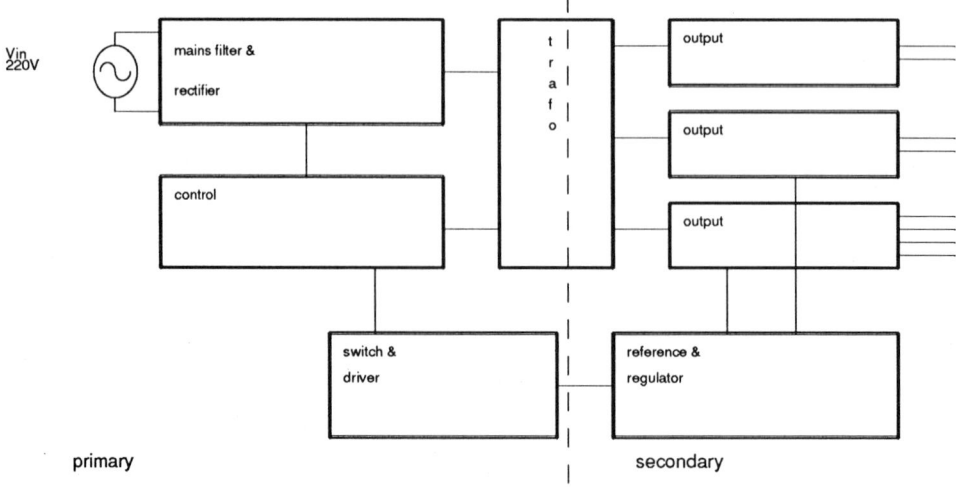

Figure A.1: Blockstructure circuit A

Block	Function
Mains filter and rectifier	Filter spikes in mains signal to power supply and filter signals coming from supply to mains
Transformer	Store energy, isolate primary and secondary side
Output sections	Rectify pulses from transformer to required DC voltage level
Reference and regulator	Regulate energy coming from primary side of the circuit to the demands on the load side of the circuit
Control section	Main oscillation circuit, performs timing of the switch function
Driver and switch	HV switch transistor and circuitry to adapt switch signal (from control circuit) to driver signals suited for the main switch transistor

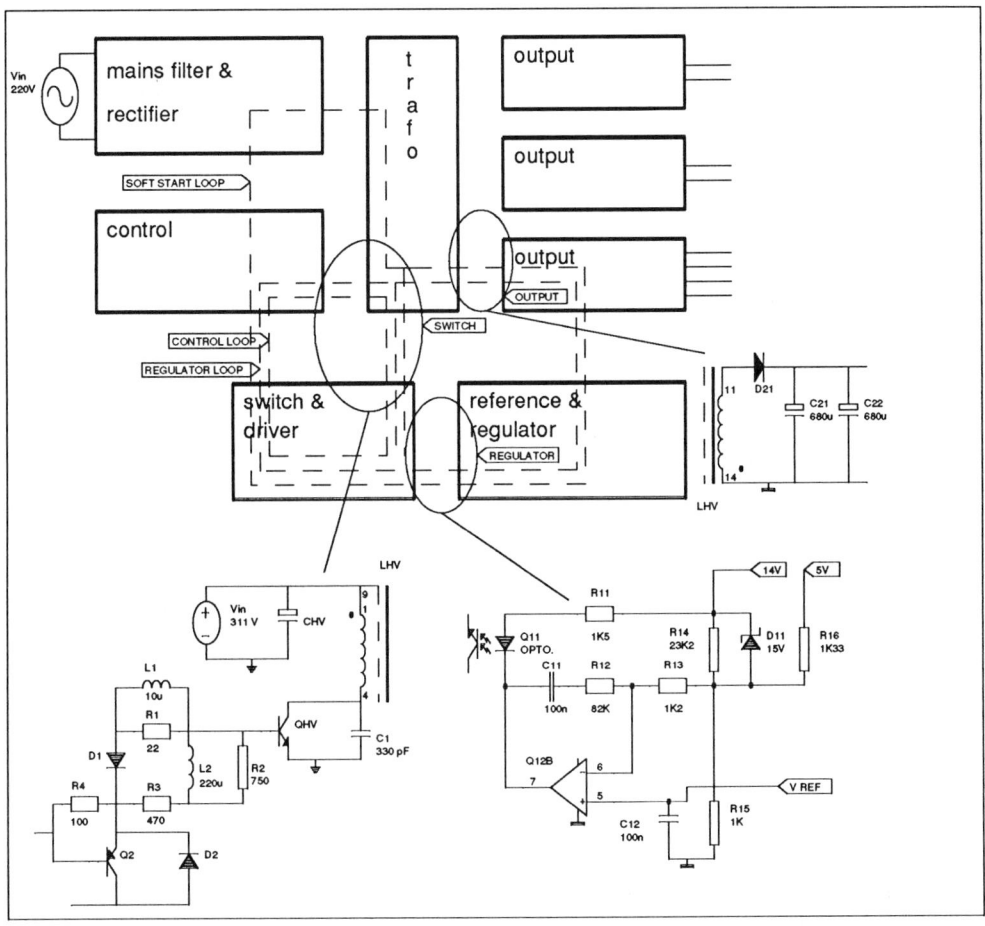

Figure A.2: Internal structure of circuit A

Figure A.2 shows, more in detail, the functional most important parts of the circuit A power supply. The most important parts of this circuit are the start-up circuit, the oscillation circuit and the regulator loop.

Simplified, the function of the power supply is to send energy packages from the primary side of the circuit to the secondary side of the circuit. For this purpose a high-voltage switching transistor activates a current through the primary inductor of the mains transformer. This current causes storage of energy in the transformer core. At the moment a certain energy level is reached (checked by using the oscillator circuit) the transistor switches off. In the now following negative phase of the oscillation cycle the stored energy flows through the secondary inductors of the mains transformer via a rectifier circuit to the load. The actual secondary voltage determines the moment the transistor switches on for a next oscillation cycle.[*] See figures A.3 and A.4.

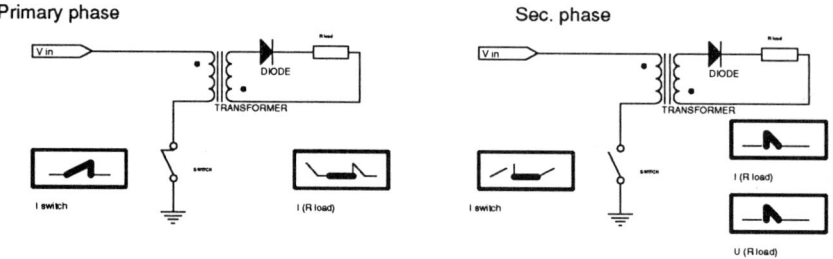

Figure A.3: Switch function

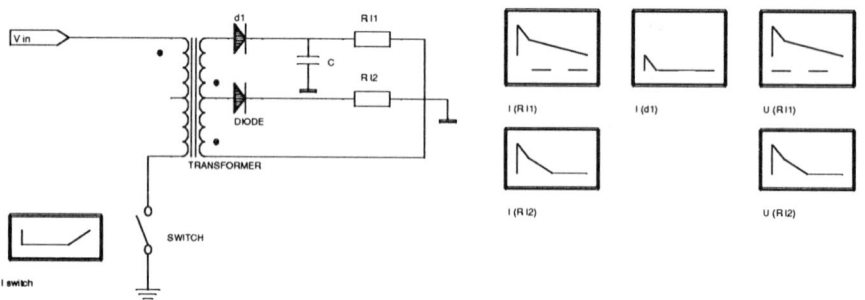

Figure A.4: Rectify function

_____

[*]   For a more detailed discussion of the functional aspects of switching power supplies: please refer to [Mar84]

Three parts of this circuit are discussed in detail. See figure A.2. The reason that these parts are discussed in detail is that either they contain traditional reliability bottlenecks or that there are considerable differences between the actual failure rates and the failure rates obtained by means of standard failure rate prediction methods.

The first part discussed in detail consists of a driver circuit, the actual switch function (mains transformer and high-voltage transistor). The second part is the reference/regulator circuit. The third part discussed in detail is closely connected to the previous part; one of the secondary rectifier circuits. The complete circuit is discussed in a separate report [Bro88].

A.4.2. *Components used (reliability aspects)*

Before it is possible to give a reliability prediction for a certain circuit according to one of the standard reliability prediction methods, it is necessary to know some details of the internal structure of the used components. [*] The table below gives these structural details for those parts of circuit A which are discussed in detail.

Code	Device type / structure
QHV	Bipolar high-speed high-voltage silicon diffused glass passivated npn power (~20W) transistor
Q2	General purpose medium power (~1W) pnp transistor
Q11	High-voltage open base optocoupler
Q12	General purpose operational amplifier
D1, D2	General purpose high-speed low power (500mW) diodes (metallurgically bonded)
D11	Low power voltage reference / regulator diode (zener diode)
D21	Medium power (~2W) schottky rectifier diode
C1	Ceramic capacitor
CHV, C21, C22	Electrolytic capacitors
Other capacitors	Plastic capacitors
Resistors	Low power (0.6W) general purpose metal film helical groove resistors

---

[*]   See Chapter 2 for explanation of the used existing reliability prediction models and parameters used in these models

Code	Device type / structure
LHV	Encapsulated aluminium foil transformer
Other inductors	Low power encapsulated coils

Using these data it will be possible to derive the (reliability) model parameters related to the device structure. Before it is possible to give a complete reliability prediction it will be necessary to find those model parameters relating to the application and the effective temperature of the various components. The tables below give the application related parameters for the components used in circuit A according to MIL and BT.

Device	MIL217	Parameters	Class/circuit parameter
All parts <> IC	λb	Basic failure rate	effective device (junction) temperature
QHV	π_a	Application factor	Switch
	π_r	Rating correction factor	$2\,W < P_{max} < 5W$
	π_{s2}	Voltage stress correction factor	$0.5 < \dfrac{\text{Applied } V_{ce}}{\text{Rated } V_{ceo}} < 0.6$
Q2	π_a	Application factor	Switch
	π_r	Rating correction factor	$P_{max} < 1\,W$
	π_{s2}	Voltage stress correction factor	$0.3 < \dfrac{\text{Applied } V_{ce}}{\text{Rated } V_{ceo}} < 0.4$
Q12	π_t	Thermal acceleration factor	effective device (junction) temperature
D1, D12	π_a	Application factor	Analog circuit
	π_r	Current rating correction factor	$I_{max} < 1\,A$
	π_{s2}	Reverse voltage stress correction factor	$\dfrac{\text{Applied } V_r}{\text{Rated } V_r} < 0.7$
D21	π_a	Application factor	Power rectifier
	π_r	Current rating correction factor	$1 < I_{max} < 3\,A$
	π_{s2}	Reverse voltage stress correction factor	$\dfrac{\text{Applied } V_r}{\text{Rated } V_r} < 0.7$
LHV	T_{hs}	Hot spot temperature	Internal power loss,

Device	BT	Parameters	Class/circuit parameter
Q11, Q12	π_t	Thermal acceleration factor	effective device (junction) temperature

From these tables it is clear that the MIL handbook uses more detailed models (considering application related influence factors) than BT. For integrated circuits both models are equal in the number of influence factors; the difference is that the BT model uses the thermal influence of two degradation mechanisms while MIL uses one.

A.4.3. *Reliability prediction using existing prediction methods*

Both the MIL-HDBK-217 and the British Telecom handbook provide models able to predict the failure rate of the used components. Under the conditions mentioned above this will result in the following failure rate figures.

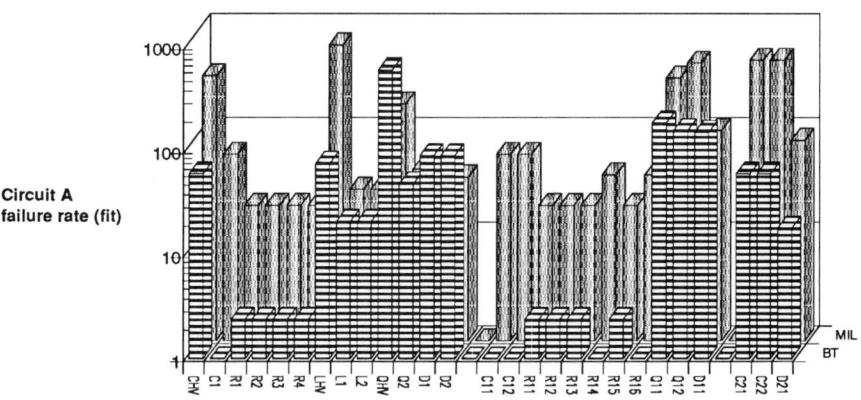

Figure A.5: Predicted failure rates circuit A

Failure rate	MIL (fails/hour)	*(fit)*	BT (fails/hour)	*(fit)*
CHV	3.5E-07	*3.5E+02*	6.0E-08	*6.0E+01*
C1	6.2E-08	*6.2E+01*	9.0E-10	*9.0E-01*

Failure rate	MIL (fails/hour)	(fit)	BT (fails/hour)	(fit)
R1	2.0E-08	2.0E+01	2.4E-09	2.4E+00
R2	2.0E-08	2.0E+01	2.4E-09	2.4E+00
R3	2.0E-08	2.0E+01	2.4E-09	2.4E+00
R4	2.0E-08	2.0E+01	2.4E-09	2.4E+00
LHV	7.0E-07	7.0E+02	7.5E-08	7.5E+01
L1	2.8E-08	2.8E+01	2.1E-08	2.1E+01
L2	2.8E-08	2.8E+01	2.1E-08	2.1E+01
QHV	1.9E-07	1.9E+02	6.0E-07	6.0E+02
Q2	4.4E-08	4.4E+01	4.8E-08	4.8E+01
D1	3.7E-08	3.7E+01	9.0E-08	9.0E+01
D2	3.7E-08	3.7E+01	9.0E-08	9.0E+01
C11	6.2E-08	6.2E+01	9.0E-10	9.0E-01
C12	6.2E-08	6.2E+01	9.0E-10	9.0E-01
R11	2.0E-08	2.0E+01	2.4E-09	2.4E+00
R12	2.0E-08	2.0E+01	2.4E-09	2.4E+00
R13	2.0E-08	2.0E+01	2.4E-09	2.4E+00
R14	3.9E-08	3.9E+01	6.0E-10	6.0E-01
R15	2.0E-08	2.0E+01	2.4E-09	2.4E+00
R16	3.9E-08	3.9E+01	6.0E-10	6.0E-01
Q11	3.3E-07	3.3E+02	1.8E-07	1.8E+02
Q12	4.7E-07	4.7E+02	1.5E-07	1.5E+02
D11	1.0E-07	1.0E+02	1.5E-07	1.5E+02
C21	5.0E-07	5.0E+02	6.0E-08	6.0E+01
C22	5.0E-07	5.0E+02	6.0E-08	6.0E+01
D21	8.4E-08	8.4E+01	1.8E-08	1.8E+01

A.4.4. *Practical failure data*

One of the problems of obtaining actual failure rate figures is the fact that, because of the relative low failure rate figures, feedback from large amounts of circuits will be necessary before it is possible to calculate a practical failure rate with satisfactory accuracy. To acquire practical reliability figures the following data-sources were available:

— Factory tests

— Burn-in and customer simulation tests

— Field failure data

In many existing reliability models the time between production and the end of the first 300 operational hours is not considered as part of the normal useful life period of a circuit. This approach would make the first two data-sources unacceptable for practical reliability figures. To compare standard reliability prediction models and practical figures possible *only field failure data* were used. As a consequence of this approach only limited data were available for the development of "standard" practical reliability figures. As the production of circuit A started end 1987 only data obtained during the period 1988-1989 were available. Figure A.6 gives practical reliability figures for those components where adequate failure figures were available.

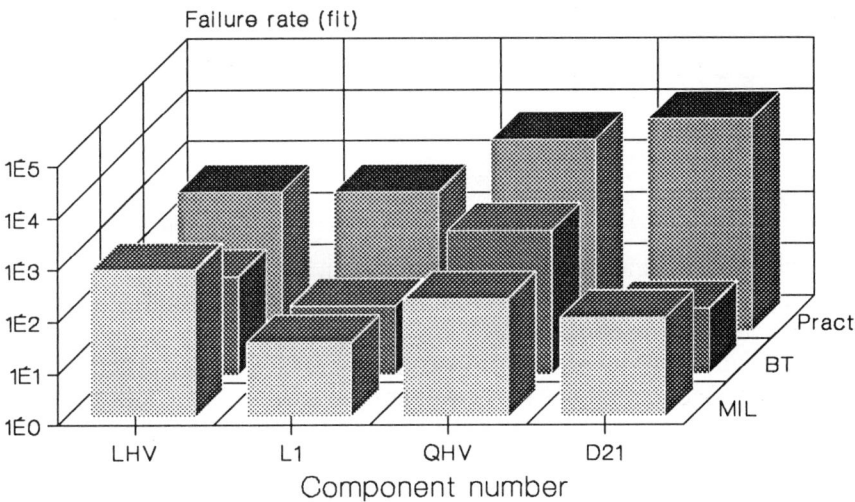

Figure A.6: Predicted/practical fail. rates circuit A

Only failure rates above a certain threshold failure rate are presented. This $\lambda_{threshold}$ represents the minimum failure rate of one failure in the entire population during the observation period. This definite minimum failure rate can be calculated using the following formula:

$$\lambda \gg \lambda_{threshold} \tag{A.1}$$

$$\lambda_{threshold} = \frac{1}{N.t} = 1.1 \times 10^{-10} \text{ (failures/hour)} \tag{A.1}$$

N: Mean number of systems available for evaluation; 10^6

t: Number of (operational[*]) hours / year; $24 * 365 = 8760$

---

[*] Most parts of a video recorder will be operational for an average period of two hours per day. In this respect the power supply forms an exception. In the current generation video recorders there are a number of functions (such as timer circuit ea.) full-time operational. As a consequence the power supply will be operational for 24 hours / day.

This number is a definite number based on the expectation of one failure per year over an observation period of one year. In practice no failure rate figures under 10^{-9} failures/hour (or 1 fit) will be given. The table below gives predicted and practical failure rates for those components where practical figures are available.

	MIL	(fit)	BT	(fit)	Practical	(fit)
LHV	7.0E-07	7.0E+02	7.5E-08	7.5E+01	5.0E-07	5.0E+02
L1	2.8E-08	2.8E+01	2.1E-08	2.1E+01	5.0E-07	5.0E+02
QHV	1.9E-07	1.9E+02	6.0E-07	6.0E+02	5.0E-06	5.0E+03
D21	8.4E-08	8.4E+01	1.8E-08	1.8E+01	1.2E-05	1.2E+04

A.5. Circuit B

Figure A.7 shows the functional structure of circuit B. Circuit B is a self oscillating power supply, similar to the power supply presented as circuit A, intended for use in video cassette recorders (VCRs). The power supply is designed to be able to deliver about 40 W output power. Circuit B is in many practical applications the successor of circuit A. This power supply is intended to fulfil the following functional demands:

Parameter	Functional demands	
Input voltage	170 to 265 V rms, 45 to 65 Hz	
Output voltage	5.2V	280 - 1200 mA
	5.2V	1200mA
	14.5V	120 - 500mA
	14.5V	40 - 850mA 1600 mA (<5sec.)
Output voltage (cont.)	33.5V	0 - 4 mA
	-29.0V	15 - 120 mA
	3.8V ~	100 - 140 mA~
Operating frequency	20 - 120 kHz	
Efficiency	78 %	

The functional outline of the circuit will be presented in the following paragraph. Only where functional aspects are important for the reliability of (parts of) the circuits they will be discussed in detail.

A.5.1. *Functional structure circuit B*

Figure A.7 shows the rough structure of the power supply. The lines in figure A.7 indicate the connections between the various modules. The structure of circuit B is quite similar to circuit A. This power supply is the successor of circuit A. Remarkable difference between both circuits is the regulator loop which is in this case located on the secondary side of the circuit.

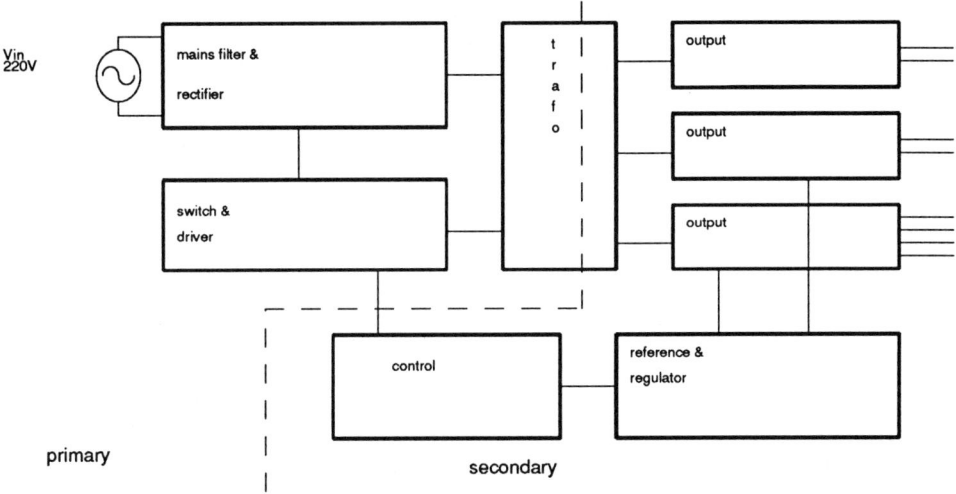

Figure A.7: Blockstructure circuit B

Figure A.8 shows, more in detail, the functionally most important parts of the circuit B power supply. These are the start-up circuit, the oscillation circuit and the regulator loop.

A.5.2. *Components used (reliability aspects)*

Before it is possible to give a reliability prediction for a certain circuit according to one of the standard reliability prediction methods it is necessary to know some details of the internal structure of the used components. The table below gives these structural details for those parts of circuit B which are discussed in detail.

Code	Device type / structure
QHV	Bipolar high-speed high-voltage silicon diffused glass passivated npn power (~20W) transistor

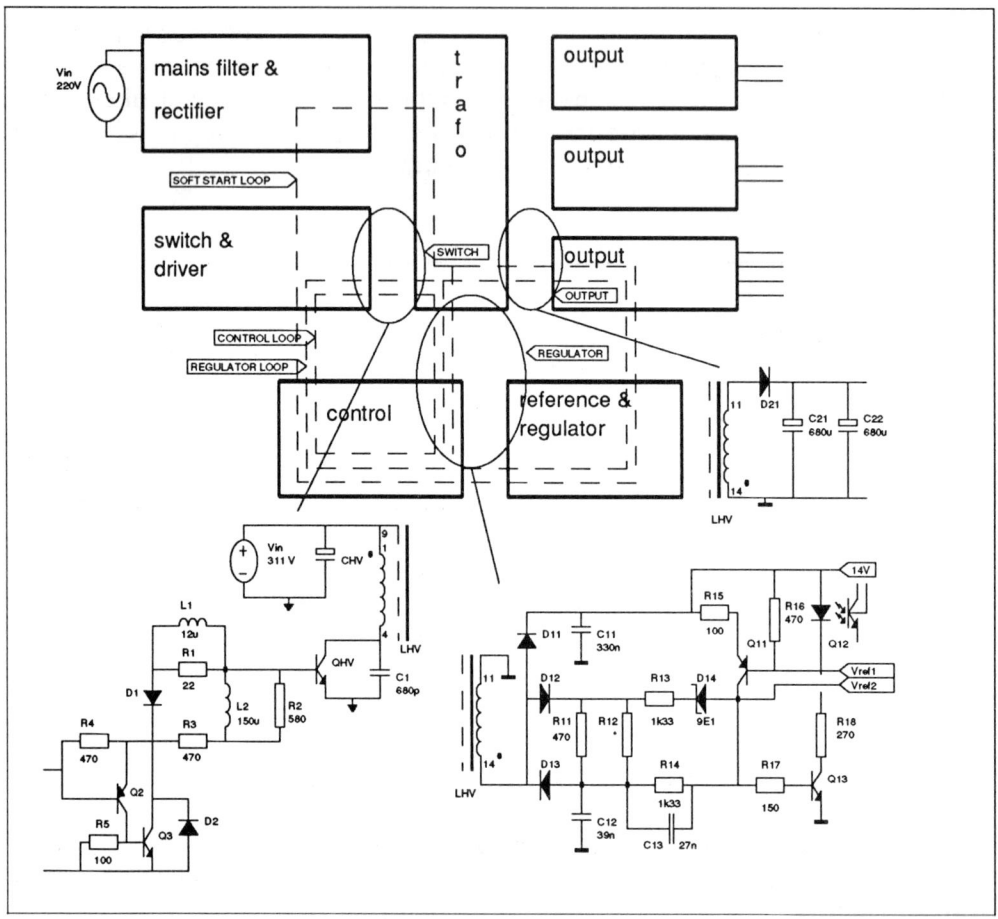

Figure A.8: Internal structure of circuit B

Code	Device type / structure
Q2, Q11	General purpose medium power (~1W) pnp transistor
Q3	General purpose medium power (~1W) npn transistor
Q12	High-voltage optocoupler
D1, D2, D11, D12, D13,	General purpose high-speed low power (500 mW) diodes (metallurgically bonded)
D14	Low power voltage reference / regulator diode (Zener diode)
D21	Medium power (~2 W) Schottky rectifier diode
C1	Ceramic capacitor

Code	Device type / structure
CHV, C21, C22	Electrolytic capacitors
Other capacitors	Plastic capacitors
Resistors	Low power (0.6 W) general purpose metal film helical groove resistors
LHV	Encapsulated aluminium foil transformer
Other inductors	Low power encapsulated coils

As discussed with circuit A these data make it possible to derive the (reliability) model parameters related to the device structure. Before it is possible to give a complete reliability prediction it will be necessary to find those model parameters relating to the application and the effective temperature of the various components. The tables below give the application related parameters for the components used in circuit B according to MIL and BT. [*]

Device	MIL Handbook	Parameters	Class/circuit parameter
All parts <> IC	λ_b	Basic failure rate	effective device (junction) temperature
QHV	π_a	Application factor	Switch
	π_r	Rating correction factor	$5 < P_{max} < 20$ W
	π_{s2}	Voltage stress correction factor	$0.5 < \dfrac{Applied\ Vce}{Rated\ Vceo} < 0.6$
Q2	π_a	Application factor	Switch
	π_r	Rating correction factor	$P_{max} < 1$ W
	π_{s2}	Voltage stress correction factor	$0.3 < \dfrac{Applied\ V_{ce}}{Rated\ V_{ceo}} < 0.4$
D1, D2, D11, D12, D13	π_a	Application factor	Analog circuit
	π_r	Current rating correction factor	$I_{max} < 1$ A
	π_{s2}	Reverse voltage stress correction factor	$\dfrac{Applied\ V_r}{Rated\ V_r} < 0.7$

---

[*]   See Chapter 2 for a more detailed discussion of the used MIL and BT reliability models.

Device	MIL Hand-book	Parameters	Class/circuit parameter
D23	π_a	Application factor	Power rectifier
	π_r	Current rating correction factor	$1 < I_{max} < 3$ A
	π_{s2}	Reverse voltage stress correction factor	$\dfrac{\text{Applied } V_r}{\text{Rated } V_r} < 0.7$
LHV	T_{hs}	Hot spot temperature	Internal power loss,

Device	BT Hand-book	Parameters	Class/circuit parameter
Q12	π_t	Thermal acceleration factor	Effective device (junction) temperature

A.5.3. *Reliability prediction using existing prediction methods*

Both the MIL-HDBK-217 and the British Telecom handbook provide models able to predict the failure rate of the used components. Under the conditions mentioned above this will result in the following failure rate figures:

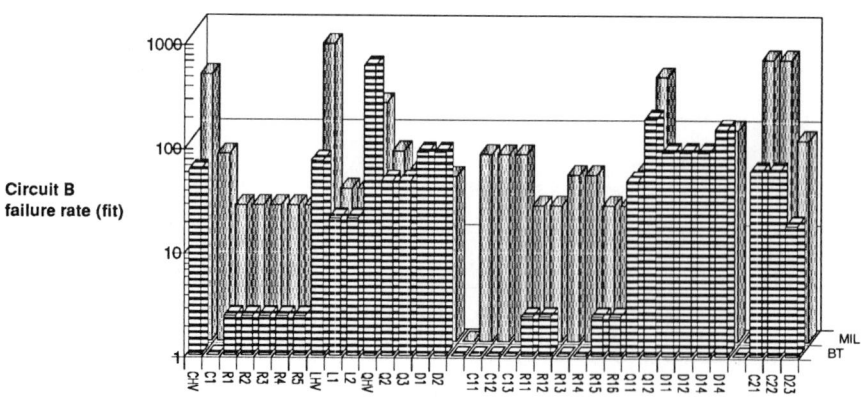

Figure A.9: Predicted failure rates circuit B

Failure rate	MIL (fails/hour)	(fit)	BT (fails/hour)	(fit)
CHV	3.5E-07	3.5E+02	6.0E-08	6.0E+01
C1	6.2E-08	6.2E+01	9.0E-10	9.0E-01
R1	2.0E-08	2.0E+01	2.4E-09	2.4E+00
R2	2.0E-08	2.0E+01	2.4E-09	2.4E+00
R3	2.0E-08	2.0E+01	2.4E-09	2.4E+00
R4	2.0E-08	2.0E+01	2.4E-09	2.4E+00
R5	2.0E-08	2.0E+01	2.4E-09	2.4E+00
LHV	7.0E-07	7.0E+02	7.5E-08	7.5E+01
L1	2.8E-08	2.8E+01	2.1E-08	2.1E+01
L2	2.8E-08	2.8E+01	2.1E-08	2.1E+01
QHV	1.9E-07	1.9E+02	6.0E-07	6.0E+02
Q2	6.6E-08	6.6E+01	4.8E-08	4.8E+01
Q3	4.4E-08	4.4E+01	4.8E-08	4.8E+01
D1	3.7E-08	3.7E+01	9.0E-08	9.0E+01
D2	3.7E-08	3.7E+01	9.0E-08	9.0E+01
C11	6.2E-08	6.2E+01	9.0E-10	9.0E-01
C12	6.2E-08	6.2E+01	9.0E-10	9.0E-01
C13	6.2E-08	6.2E+01	9.0E-10	9.0E-01
R11	2.0E-08	2.0E+01	2.4E-09	2.4E+00
R12	2.0E-08	2.0E+01	2.4E-09	2.4E+00
R13	3.9E-08	3.9E+01	6.0E-10	6.0E-01
R14	3.9E-08	3.9E+01	6.0E-10	6.0E-01
R15	2.0E-08	2.0E+01	2.4E-09	2.4E+00
R16	2.0E-08	2.0E+01	2.4E-09	2.4E+00
Q11	4.4E-08	4.4E+01	4.8E-08	4.8E+01
Q12	3.3E-07	3.3E+02	1.8E-07	1.8E+02
D11	3.7E-08	3.7E+01	9.0E-08	9.0E+01
D12	3.7E-08	3.7E+01	9.0E-08	9.0E+01
D14	3.7E-08	3.7E+01	9.0E-08	9.0E+01
D14	1.0E-07	1.0E+02	1.5E-07	1.5E+02
C21	5.0E-07	5.0E+02	6.0E-08	6.0E+01
C22	5.0E-07	5.0E+02	6.0E-08	6.0E+01
D23	8.4E-08	8.4E+01	1.8E-08	1.8E+01

A.5.4. *Practical failure data*

Obtaining practical reliability figures for circuit B the same problems areise as discussed with circuit A. For that reason only failure rates with a value above 10^{-9} failures/hour circuit B are given.

	MIL	(fit)	BT	(fit)	Practical	(fit)
LHV	7.0E-07	7.0E+02	7.5E-08	7.5E+01	5.7E-07	5.7E+02
L1	2.8E-08	2.8E+01	2.1E-08	2.1E+01	5.7E-07	5.7E+02
QHV	1.9E-07	1.9E+02	6.0E-07	6.0E+02	5.0E-06	5.0E+03
Q11	4.4E-08	4.4E+01	4.8E-08	4.8E+01	7.5E-06	7.5E+03

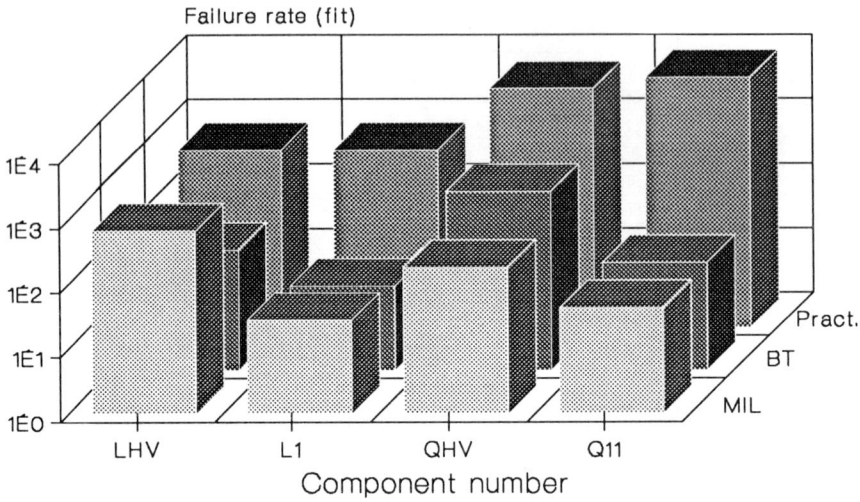

Figure A.10: Predicted/practical fail. rates circuit B

A.6. **Circuit C**

The main function of circuit C is rotating the head disc of a video cassette recorder at well defined speed which depends on the operation mode of the recorder. Information considering the current position of the head disc is also processed in this circuit. See figure A.11 for the block-structure of this circuit.

A.6.1. *Function*

The most important parts in circuit C are the head disc motor and the motor driver IC. As the function of the circuit is driving and controlling the motor (speed and position) of a headscanner most functional details will be related to the motor, the driver circuits and the feedback-and control circuits. See figure A.12 for a more detailed circuit diagram.

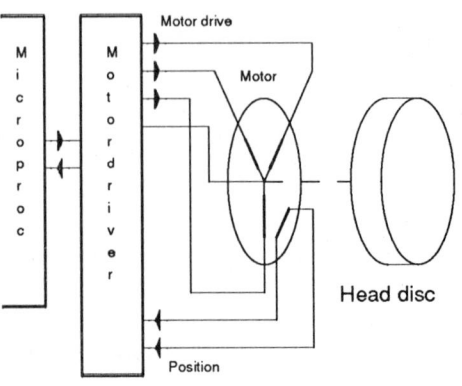

Figure A.11: Blockstructure circuit C

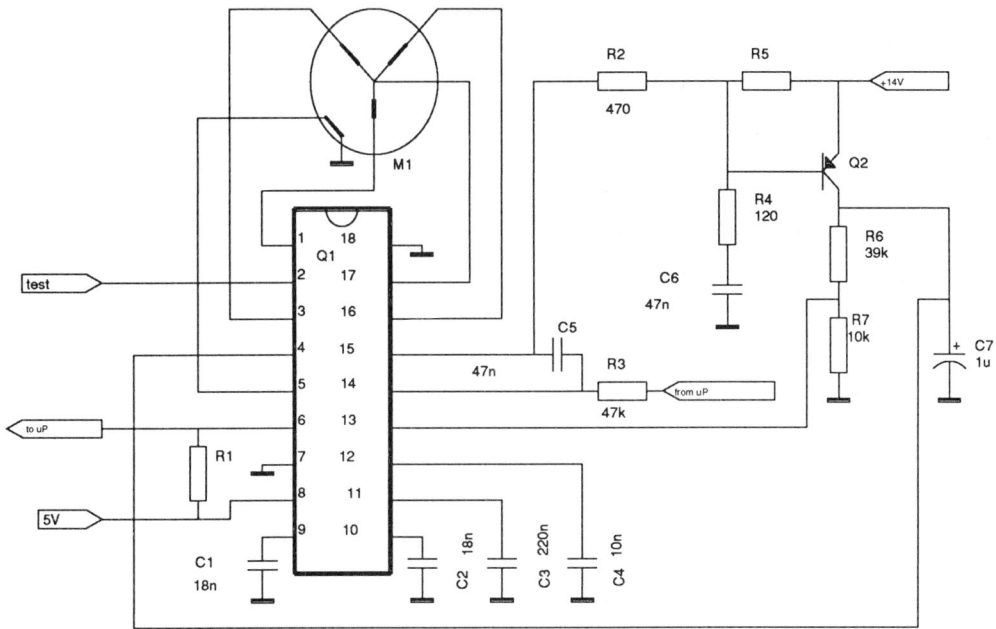

Figure A.12: Schematic structure of circuit C

The motor is a three -phase star motor. A major part of the circuitry is found in one integrated circuit. See figure A.13. This motor drive IC is a mixed analog / digital integrated circuit. It is developed especially as a full-wave drive unit for VHS video cassette recorder motors.

A.6.1.1 *The motor drive IC*

The Philips motor drive integrated circuit is a three phase motor drive, which is developed for sensorless, full wave and high current star and triangle connected motors. Internally the commutation logic senses the rotor position using the motor EMF, to drive the output stages. Thermally the driver IC is protected against high currents. The circuit is suited as a drive unit for VHS video cassette recorder motors.

The following parts of the motor drive IC can be determined:

Block	Function
Operational trans-conductance am-plifier	Amplifier is independent of the rest of the IC, is often used in typical motor drive applications. (analog)
Power output stages	Switching the motor current (analog, power)

Block	Function
Thermal protection circuit	Protecting the circuit against overpower by means of limiting/controling the power stages (analog)
Bandgap reference circuit	Reference circuit, used by commutation logic. (analog)
EMF-comparators	Feedback from motor circuit used to obtain the current motor status (analog)
Commutation logic	Control of the motor drive switches (digital)
Start-up oscillation circuit	(digital)
Control and timing logic	(digital)

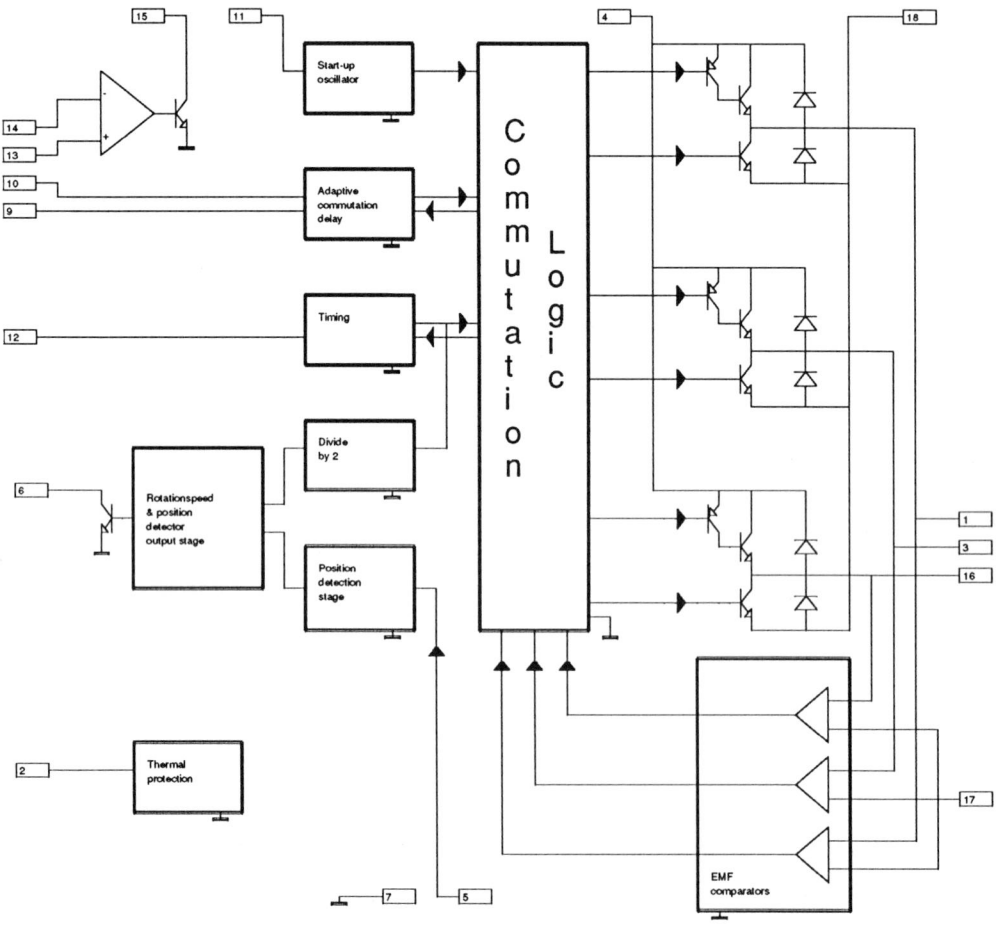

Figure A.13: Internal structure motor drive IC

The output stages contain flyback diodes because of the high inductive loading.

A.6.1.2 *Function of the motordrive IC in circuit C*

Full-wave driving of a three phase motor requires three push-pull output stages In each of the six possible states are two outputs active, one sourcing and one sinking current. See figure A.14. The third output shows a high impedance to the motor. On that output it is possible to measure the motor EMF on the corresponding motor coil. That is done by the EMF comparator on each output. The commutation logic provides the proper control for the output transistors and the selection of the right EMF comparator. The detected zero- crossing in the motor EMF is used to calculate the right moment for the next output pulse, that is the change to the next output state. The delay is calculated, depending on the load of the motor, by the adaptive commutation delay block [Lee88].

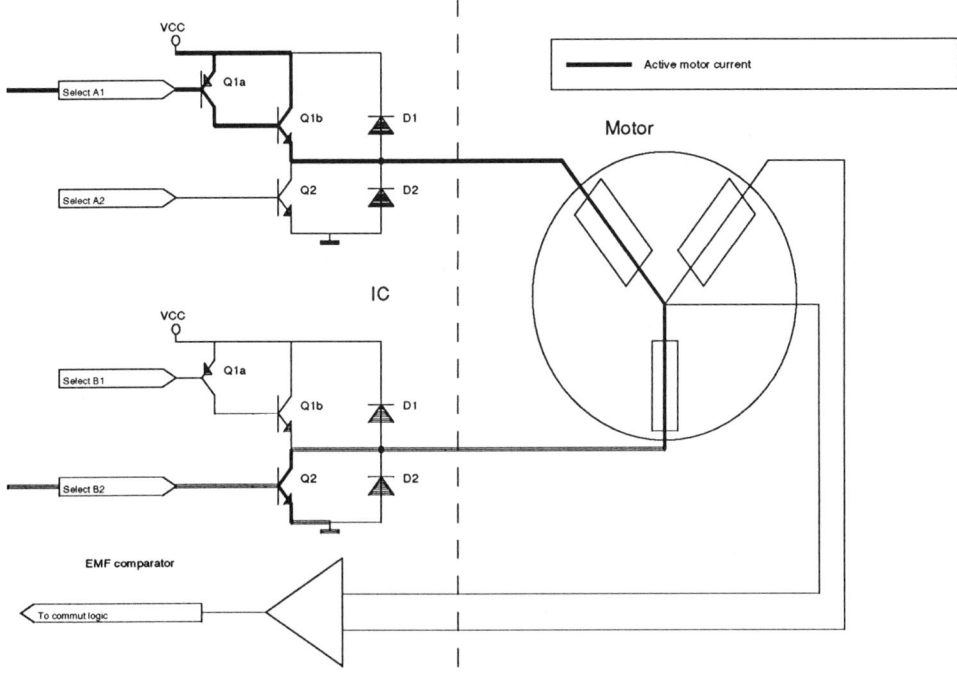

Figure A.14: Function circuit C

A.6.2. *Components used (reliability aspects)*

Before it is possible to give a reliability prediction for a certain circuit according to one of the standard reliability prediction methods it is necessary to know some details of the internal structure of the used components. The table below depicts these structural details for those parts of circuit C which are discussed in detail.

Code	Device type / structure
Q1	Mixed digital/analog bipolar integrated circuit.
Q2	General purpose medium power (~2W) pnp transistor
C7	Electrolytic capacitor
Other capacitors	Plastic capacitors
Resistors	Low power (0.6W) general purpose metal film helical groove resistors
M1	Three phase star motor to be used as headscanner drive in VCRs

As discussed with circuit A these data make it possible to derive the (reliability) model parameters related to the device structure. Before it is possible to give a complete reliability prediction it will be necessary to find those model parameters relating to the application and the effective temperature of the various components. The tables below list the application related parameters for the components used in circuit C according to MIL and BT. [*]

Device	MIL	Parameters	Class/circuit parameter
All parts <> IC & Motor	λ_b	Basic failure rate	Effective device (junction) temperature
Q1	π_t	Thermal acceleration factor	Effective device (junction) temperature
Q2	π_a	Application factor	Analog
	π_r	Rating correction factor	$1 < P_{max} < 5$ W
	π_{s2}	Voltage stress correction factor	$0.3 < \dfrac{\text{Applied } V_{ce}}{\text{Rated } V_{ceo}} < 0.4$

Device	BT	Parameters	Class/circuit parameter
Q1	π_t	Thermal acceleration factor	Effective device (junction) temperature

A.6.3. *Reliability prediction using existing prediction methods*

Both the MIL-HDBK-217 and the British telecom handbook provide models able to predict the failure rate of most of the used components. The only problem in this respect is reliability prediction of the motor according to MIL. This reliability

---

[*]   See Chapter 2 for a more detailed discussion of the used MIL and BT reliability models.

prediction handbook uses a model which is fundamentally different from the models of other components. Therefore this model will be discussed in more detail.

A.6.3.1 *The MIL-HDBK-217 motor model*

One of the more difficult aspects of predicting the reliability of a VCR motor is the problem that a motor is only partially an electronic component. A considerable part of the motor's behaviour depends also on the mechanical properties of the motor. This chapter will concentrate, however, on the electrical properties of the motor circuit. For the reliability prediction only the electrical reliability of the circuit will be discussed. Only as part of the discussion for the MIL reliability prediction model, mechanical properties of the motor will be taken into account. MIL-HDBK-217E uses the following reliability model for electro-mechanical components such as motors:

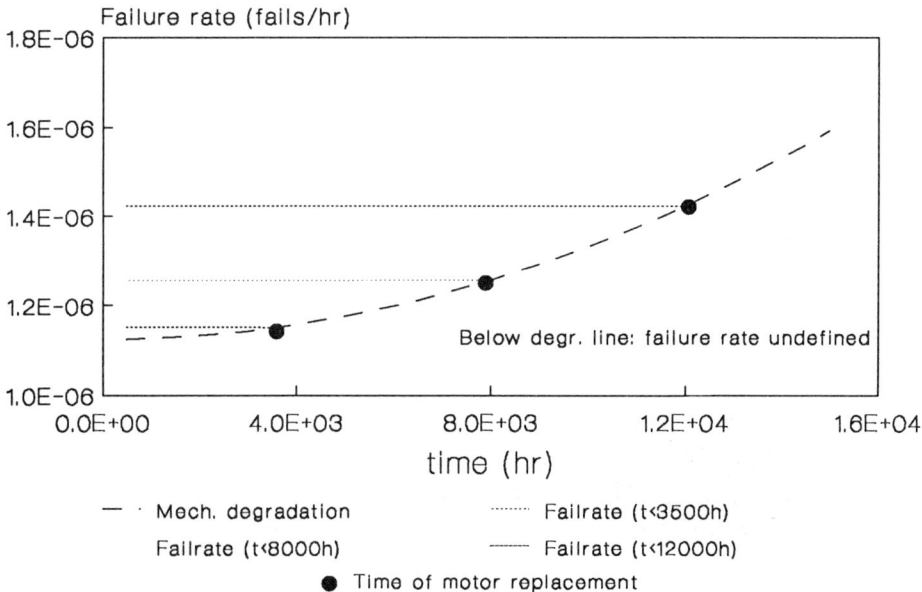

Figure A.15: MIL failure rate model motors

$$\lambda_{\text{motor}} = \frac{t^2}{\alpha_b^3} + \frac{1}{\alpha_w} \qquad\text{(A.2)}$$

— t: motor operating time. (total elapsed time before replacement)

— α_b: Weibull characteristic of bearing life constant (mechanical)

— α_w: Characteristic life constant of winding (electrical)

In this case the time-dependent failure rate is approximated by its (worst- case) failure rate, that is the failure rate of the mechanical parts of the motor at the end of its lifetime, at the moment that motor replacement is expected. This (worst-case) failure rate is used, assuming a constant failure rate. Of the discussed reliability handbooks only MIL uses "total-life time" models for electro-mechanical components; constant failure rate models where the failure rate level depends on the mechanical aspects of the component at the expected time of part-replacement. See also figure A.15. As the mechanical aspects of reliability are not within the scope of this book mechanical aspects of reliability will not be discussed here in further detail.

A.6.3.2 *Reliability figures used components*

Using the presented data it will be possible to make a reliability prediction for all components used in circuit C. Figure A.16 and the table below will present the result of these calculations. In contrary to the circuits A and B it is not (yet) possible to present practical failure rate figures of circuit C. As the large series production of circuit C will start early 1990 it will probably take until the end of 1990 before practical failure rate figures are available.

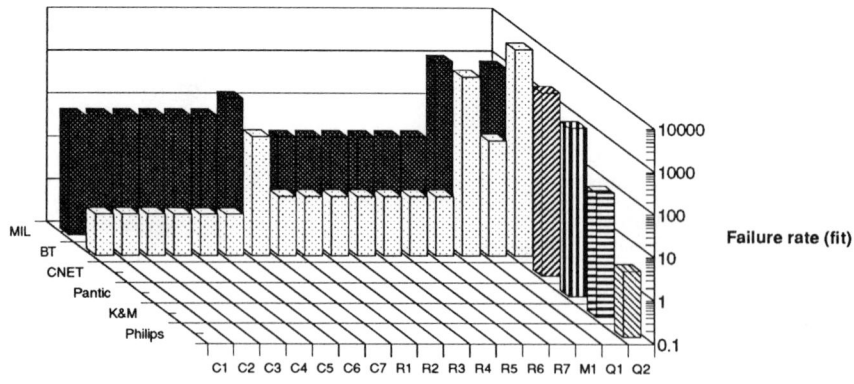

Figure A.16: Predicted failure rate circuit C

Failure rate	MIL (fails/hour)	(fit)	BT (fails/hour)	(fit)
C1	6.2E-08	6.2E+01	9.0E-10	9.0E-01
C2	6.2E-08	6.2E+01	9.0E-10	9.0E-01
C3	6.2E-08	6.2E+01	9.0E-10	9.0E-01
C4	6.2E-08	6.2E+01	9.0E-10	9.0E-01
C5	6.2E-08	6.2E+01	9.0E-10	9.0E-01
C6	6.2E-08	6.2E+01	9.0E-10	9.0E-01
C7	1.6E-07	1.6E+02	6.0E-08	6.0E+01
R1	2.0E-08	2.0E+01	2.4E-09	2.4E+00
R2	2.0E-08	2.0E+01	2.4E-09	2.4E+00
R3	2.0E-08	2.0E+01	2.4E-09	2.4E+00
R4	2.0E-08	2.0E+01	2.4E-09	2.4E+00
R5	2.0E-08	2.0E+01	2.4E-09	2.4E+00
R6	2.0E-08	2.0E+01	2.4E-09	2.4E+00
R7	2.0E-08	2.0E+01	2.4E-09	2.4E+00
Q1	6.7E-08	6.7E+01	4.8E-08	4.8E+01
Q2	8.4E-07	8.4E+02	6.6E-06	6.6E+03
M1	1.2E-06	1.2E+03	1.5E-06	1.5E+03

Appendix B
Failure Mechanisms in
Simple Components

Chapter 4 presented a number of failure mechanisms, occurring in practical components, and derived practical susceptibility models for two semiconductors. This chapter finally presents susceptibility models for these components, based on a combination of theoretical models and practical measurements. The disadvantage of Chapter 4 is that only models for two components are given. The purpose of this appendix is to present some guidelines for the development of susceptibility models for other components. The presented models and failure mechanisms are by no means complete and not suited for direct use. The presented guidelines are intended as a platform for further discussions and further research.

B.1. Resistive components

The best known common property of resistive components is that a current through the component will cause a certain voltage drop. The relation between current and voltage drop is expressed in the well-known ohmic law:

$$V = R \times I \tag{B.1}$$

where R is the resistance of the component.

The exact value of this resistance depends strongly on the material used in the component and the structure of the component.

B.1.1. *General failure mechanisms*

B.1.1.1 *Power overstress*

One of the best known failure mechanisms for resistive components is the burning of the resistive material, finally resulting in an open circuit. Due to (local) hot spots in the resistor the temperature of the material might reach a value where irreversible degradation effects occur. There are two important influences on the temperature of a resistive device:

— Environmental temperature

— Temperature rise caused by internal power dissipation

The exact (local) temperature within the device depends strongly of the nature of the power dissipation (DC or pulsed) and on the thermal resistance of the device to ambient. See also Chapter 4. For DC operation the main susceptibility limit is usually:

— Device temperature

In those cases where a resistive device does not have a homogeneous structure it is possible that differences exist in local power dissipation. In this case the main susceptibility limit becomes:

— Device hot-spot temperature

B.1.1.2 *Pulse power effects*

As discussed in Chapter 4 there is a considerable difference in thermal effects between DC power dissipation and pulse power dissipation. In case of pulse

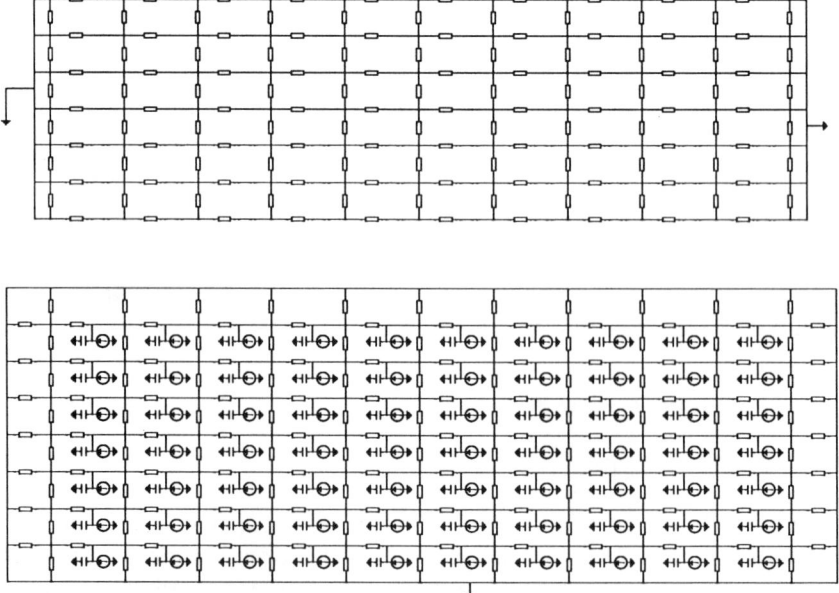

Figure B.1: Electrical and thermal model resistor

power operation the resulting temperature of the device strongly depends on the thermal resistance and thermal capacitances of the device. See figure B.1 for the electrical and thermal model of a practical resistor and figure B.2 for a detailed part of this resistor.

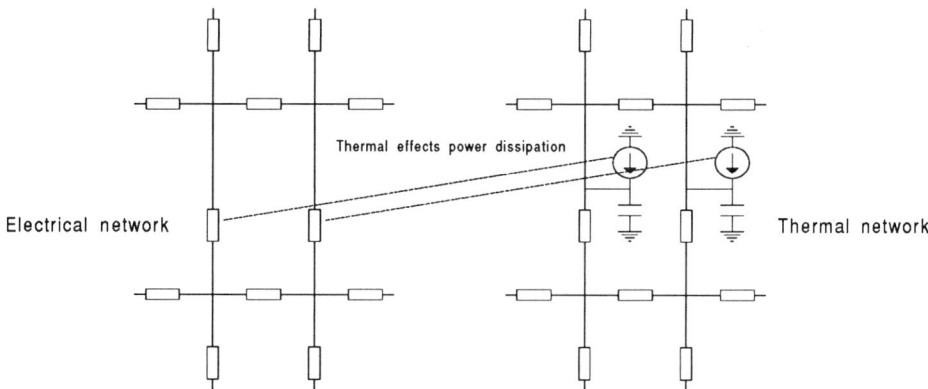

Figure B.2: Detail of figure B.1

In those cases where the resistor is stressed using pulses, the temperature on a certain point of the resistor will depend strongly on the transient effects of the pulses, both in the electrical domain (power dissipation) and in the thermal domain (thermal resistances and capacitances). See figure B.3.

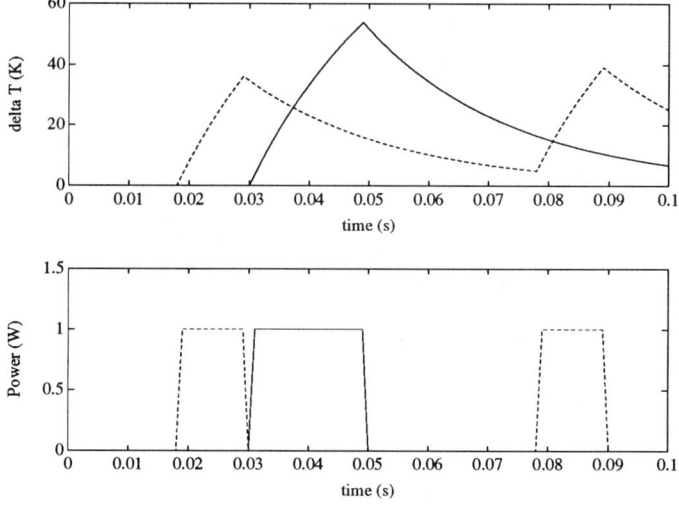

Figure B.3: Electrical and thermal transient effects

Also for pulse-power operation of a device the spot temperature of the device should not exceed a certain maximum value. Even for homogeneous devices the spot temperature will be different within a device, depending on the nature of the applied pulses. Figure B.4 shows the thermal effect on the maximum device temperature of pulses with a time-constant less than the time-constant of the thermal network, figure B.5 shows the effect of pulses with a comparable energy but now with a time-constant comparable to the time- constant of the thermal network.

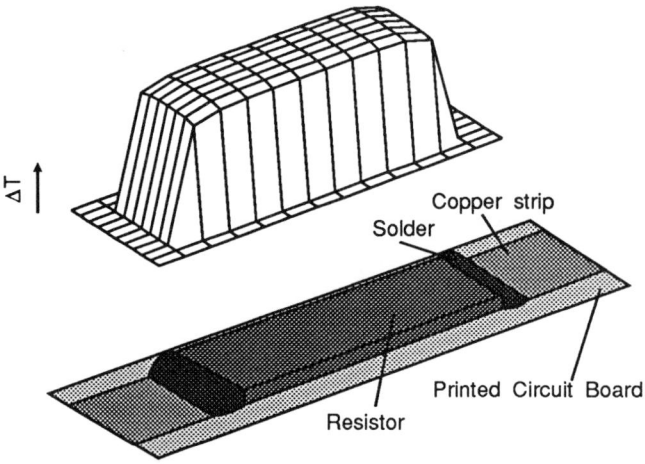

Figure B.4: Effects on Tmax of fast pulse, low dutycycle

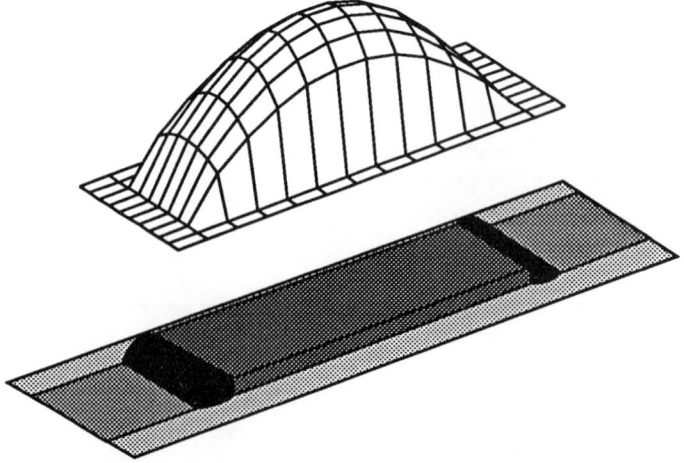

Figure B.5: Effects on Tmax of slow pulses or DC

Using the above model it is possible that, in case of pulsed operation, two factors influence the device temperature.

— Duration of the pulses

— Pulse interval

For other operation modes than pulse operation it is possible to use the same dynamic thermal model to derive device temperature also for more complicated signal form.

B.1.1.3 *Effects of inhomogeneities on power dissipation*

The previous section discussed effects on power dissipation (hot spot temperature) of D.C. and pulsed operations of a resistor. Another important cause for hot-spot formation is the occurrence of dislocations in a device causing possibilities for increased current densities. Figure B.6 shows the effect of a damage spot in a resistor on the hot-spot temperature of the device. Hot-spots are especially likely in cases of combinations of short pulses and dislocations in the material.

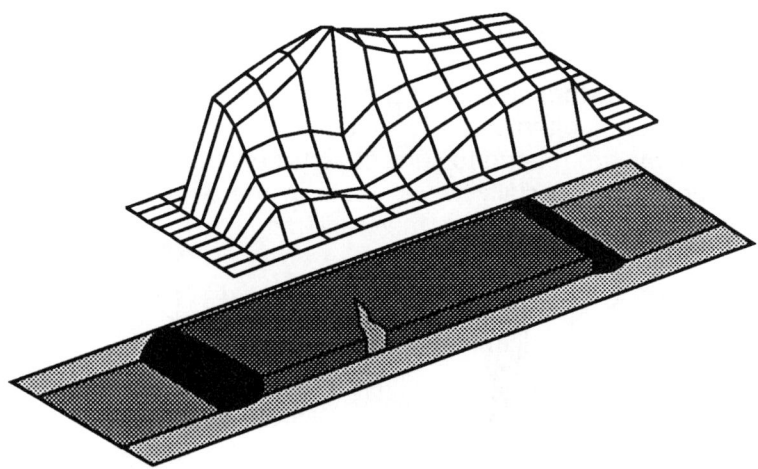

Figure B.6: Effects on Tmax of fast pulse, damaged R

B.1.1.4 *Voltage overstress*

Although the burning of a resistor is one of the best known failure mechanisms for this class of components there is yet another class of failure mechanisms, especially important for devices with high resistance values.

These are the effects related to the various forms of high-voltage breakdown. *
In those cases where a considerable electric field exists within the component
it is possible that this electric field may cause one of the high- voltage breakdown
mechanisms discussed in Chapter 4. As discussed in this chapter there are two
categories of breakdown possible which might be applicable for this category of
devices.

    — Impact ionization

    — Electron trap ionization

As mentioned in Chapter 4 of these two failure mechanisms the first is not time
dependent while the second mechanism is time dependent. Only relatively slow
changing electric fields will be able to initiate electron trap ionization. From
manufacturers databooks it is quite often possible to distinguish the area where
impact ionization dominates and the area where electron trap ionization has the
most important influence. The following sections will give two practical examples
of resistors, their failure mechanisms and the relation with the used technology.

B.1.2. *Practical components*

B.1.2.1 *Wire wound resistors*

Most wire wound resistors consist of a wire of material with a certain resistivity
wound on a cylinder of, often ceramic, material. In wire wound resistors the main
failure mechanisms are related to current densities and/or high (local) temperatures in the device. Especially weak spots in the wire material have a high
susceptibility for the various failure mechanisms.

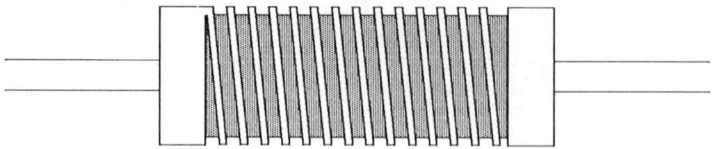

Figure B.7: Wire wound resistor

See figures B.7 and B.8 for the (simplified) internal structure of wire wound
resistors and a possible failure effect. As wire wound structures are used most
times in relatively low resistance ranges (quite often especially for high-power

---

*   Although theoretically high voltage breakdown is possible for all resistance values,
    power breakdown will usually have a dominant effect for low resistance values.

purposes) wire wound resistors are likely to be susceptible to power overstress only. This gives the following stressors for power resistors.

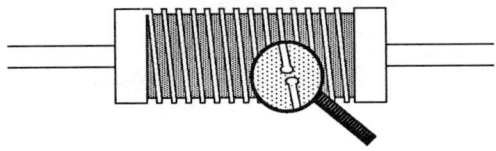

Figure B.8: Wire wound resistor failed; power stress

Failure mechanism wire wound resistors	Stressor	Related material aspects
Power overstress	Environmental tempera-ture	
	Steady state power dissi-pation (Hot spot temperature)	Thermal resistance to ambient
	Pulse power dissipation (Hot spot temperature)	Thermal resistance to ambient Thermal capacitance Inhomogeneities

B.1.2.2 *Film resistors*

Film resistors consist of a certain layer of conductive material deposited on a non-conductive substrate. See figure B.9. By means of a trimming or cutting process the deposited material is trimmed in such a way that a resistor with a certain resistance value is produced. Principal failure mechanisms are related to the current density and local temperature in the conducting layer. See also the previous sections. See figure B.10 for a possible failure effect.

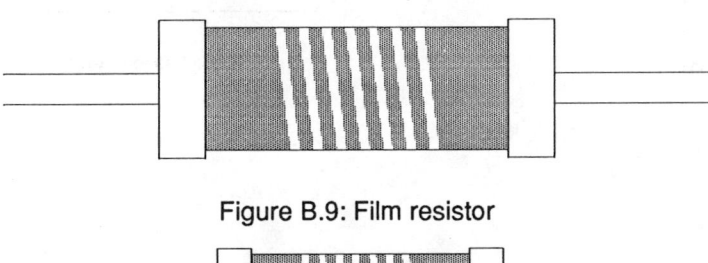

Figure B.9: Film resistor

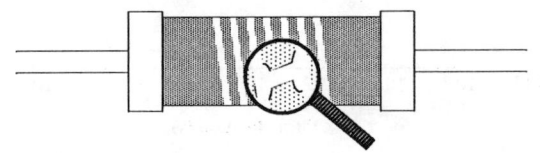

Figure B.10: Film resistor failed; power dissipation

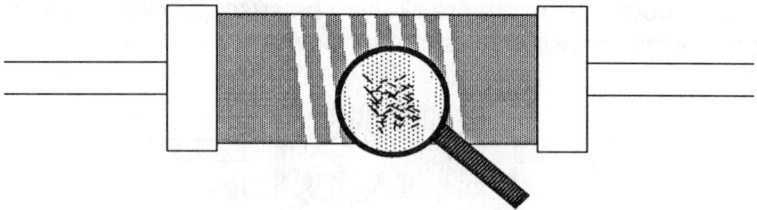

Figure B.11: Micro cracks in film resistors

Additional localized heating might cause so-called micro cracks due to differen-ces in thermal expansion coefficient of the used materials. Ultimately hot-spots may turn the resistor into an open circuit. See also figure B.11.

— Thermal expansion coefficient used materials

— High voltage breakdown characteristics used material (especially where high E fields exist)

Interesting aspect of high-voltage breakdown in film resistors is that both impact ionization and electron trap ionization can be found in these components. As electron trap ionization is a slower effect compared to impact ionization the first effect will occur especially in those cases where pulses with large pulse-widths occur. Fast pulses will not be able to activate this failure mechanism and

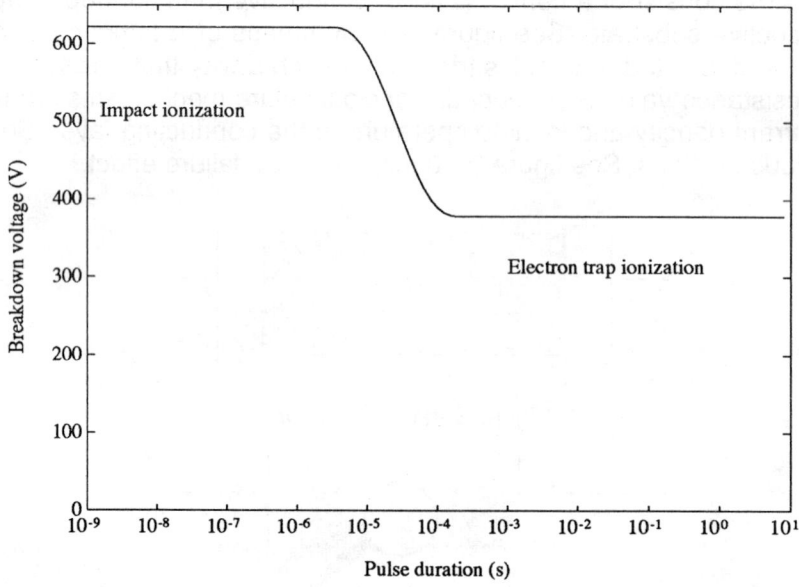

Figure B.12: Impact- and electron trap- ionization

therefore the breakdown voltage for short pulses will depend on impact ionization only. See figure B.12 for a typical relation between pulse-width and breakdown voltage. Figures like this can be found in many component manufacturers databooks.

Stressors related to the above failure mechanisms are

Failure mechanism film resistors	Stressor	Related material aspects
Power overstress		Thermal expansion coefficient used materials
	Environmental temperature	
	Steady state power dissipation (Hot spot temperature)	Thermal resistance to ambient
	Pulse power dissipation (Hot spot temperature)	Thermal resistance to ambient Thermal capacitance
High voltage breakdown (Electron trap)	Applied E field pulse-width of E field	Amount of impurities in the material used
High voltage breakdown (Impact ionization)	Applied E field, dE/dt	Electric breakdown characteristics of the material used

B.2. Capacitive components

Most capacitive components consist, in principle, of two layers of conducting material separated by means of a layer of isolating material. To limit the total area of the capacitor quite often multi layer structures are used. Relation between voltage and current for such a device can be expressed using

$$I = C.\frac{dV}{dt} \tag{B.2}$$

In practice most capacitors will not have this ideal behavior. An often used model for practical capacitors is given in figure B.13. Important parameters in this respect are:

— R_s: series resistance of the capacitor

— L_s: series inductance of the capacitor

— R_p: parallel resistance of the capacitor

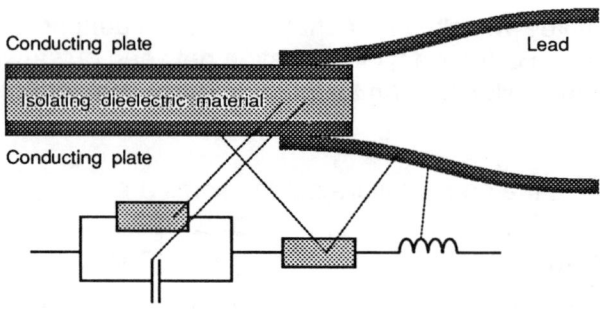

Figure B.13: Model simple capacitor

The series resistance is related to the ohmic resistance of the leads and the plates while the parallel resistance is related to the resistance of the dielectric material. The series inductance is related to the inductance of the capacitors leads and plates.

B.2.1. *General failure mechanisms*

B.2.1.1 *High-voltage breakdown*

Main failure mechanisms of simple capacitive components are related to the isolating aspects of the dielectric material. The resistivity of the isolating material might cause ohmic losses or the electric field across the capacitor might cause high voltage breakdown. The type of high voltage breakdown depends very much on the structure of the used isolating material. Most likely form of breakdown is in many cases impact breakdown. In those cases where a considerable amount of impurities exists also breakdown as effect of electron trap ionization is possible. High-voltage breakdown failure mechanisms in such a component are related to these basic material properties:

— Used material

— Internal structure

The following stressors are related to the discussed failure mechanisms

— Electric field

— Changes in electric field $\dfrac{\mathrm{d}\,E}{\mathrm{d}\,t}$ (only in case of a considerable amount of impurities)

— Temperature (significant influence mainly in cases of a considerable amount of impurities)

B.2.1.2 *Effects of inhomogeneous structures on voltage breakdown*

The previous paragraphs discussed high-voltage breakdown especially as a function of the dielectric material. The structure of the component has also an important influence on the failure probability of a device. Especially at locations within the component where high *E fields occur there is an increased probability of high voltage breakdown.* See figure B.14.

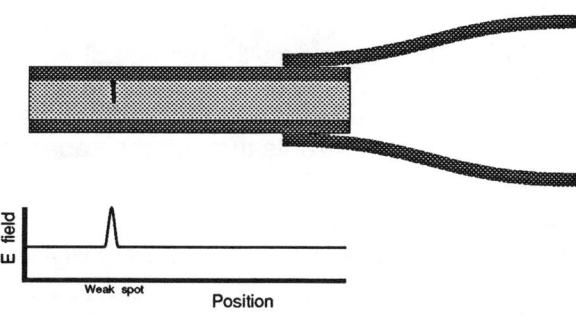

Figure B.14: Inhomogeneous plate causes E peak

B.2.1.3 *Power breakdown*

Although theoretical, ideal, capacitors do not dissipate power the non-ideal capacitor of figure B.13 shows clearly two areas where power dissipation is possible: the series resistance and the parallel resistance. The series resistance relates to the resistance of plates and leads while the parallel resistance relates to ohmic losses in the dielectric material. For many capacitors the series resistance is too small and the parallel resistance is too large to cause any serious problems. Only cases of high dE/dt will result in currents sufficiently high to cause considerable power dissipation in the series resistance (See also the paragraph on electrolytic capacitors).

For most capacitors it is possible to consider high-voltage breakdown as dominant failure mechanism.

B.2.2. **Practical components**

B.2.2.1 *Ceramic/plastic capacitors*

Many, especially small value, capacitors use ceramics or plastic as dielectric material. In those cases where the area required to achieve the wanted capacitance value becomes too large for practical application, multi layer structures are used. See figure B.15. Generally speaking this class of capacitors is only susceptible for high- voltage breakdown. (Either impact- or electron trap-

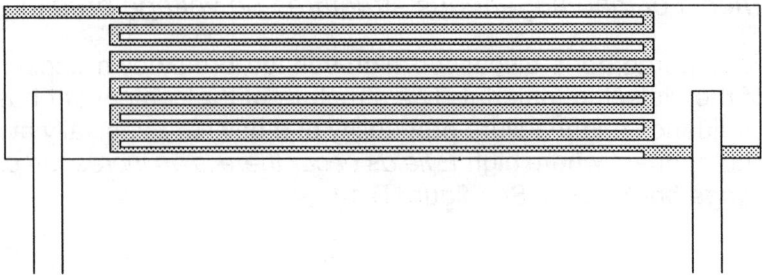

Figure B.15: Simple multi-layer capacitor

ionization) As the capacitance values for these capacitors are usually small (<1 µF) it is quite uncommon to have currents through the capacitors where effects of power dissipation are possible.

Failure mechanism film capacitors	Stressor	Related material aspects
High voltage breakdown (Electron trap)	Applied E field pulse-width of E field	Amount of impurities in the material used
High voltage breakdown (Impact ionization)	Applied E field, dE/dt	Electric breakdown characteristics of the material used

B.2.2.2 *Electrolytic capacitors*

Main difference between standard capacitors and electrolytic capacitors is the difference in used dielectric material. While other capacitors have an internal structure consisting of two plates and an isolating dielectric material in between, the internal structure of an electrolytic capacitor consists of two metal (often aluminium) foils covered with a thin layer of aluminium oxide. To achieve high capacity and prevent intolerable low breakdown voltages one of the capacitor plates is covered with a thicker layer of aluminium oxide. (As a consequence the capacitor will have different breakdown voltages for both polarities.) The space

☐ Al
▨ AlO2
▦ Electrolyt

Figure B.16: Electrolytic capacitor

between both foils is filled with solid or fixed electrolyte. The electrolyte and the two foils are conducting while the aluminium oxide is isolating. See figure B.16. Main reason for this construction is the possibility to achieve comparatively high capacitance values (often > 1mF) within a limited volume. As the volume is one of the major constraints for this component type there is a tendency to use very thin foils of material.

As a consequence it will no longer be possible to neglect some effects which have only minor influence on other capacitors. In contrast with other capacitors in the case of electrolytic capacitors not only high voltage breakdown is possible but also effects related to power dissipation. In ideal capacitors power dissipation is not possible.

In practical capacitors power dissipation might be caused by ohmic losses in the dielectric material. For electrolytic capacitors (or other capacitors with very large, thin capacitor plates) there is an additional source of power dissipation. Especially in those cases where considerably alternating electric fields exist within the capacitor there is a considerable amount of charge flowing in- and out- of the capacitor plates. Due to the comparatively high resistivity of the very thin and large aluminium foils these moving charges (currents) may cause a voltage drop between points on the capacitor plates and the terminals of the component, thus causing power dissipation. See figure B.17.

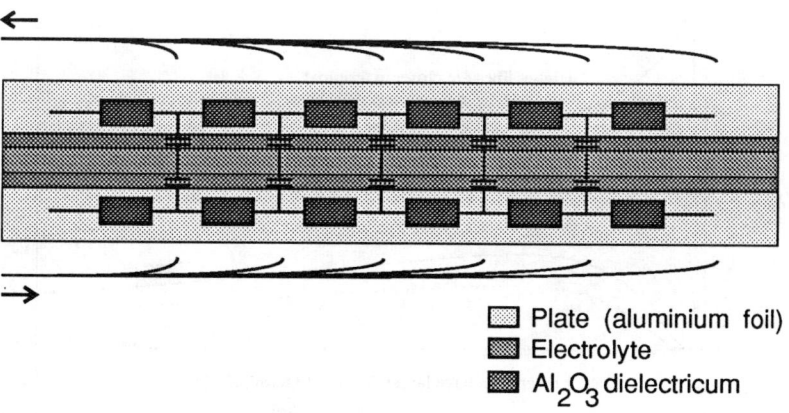

☐ Plate (aluminium foil)
▨ Electrolyte
▨ Al_2O_3 dielectricum

Figure B.17: Currents in ELCOs during pulse operation

To demonstrate the effect of local power dissipation in an electrolytic capacitor figures B.18 and B.19 show the effects of a voltage pulse applied to a capacitor.

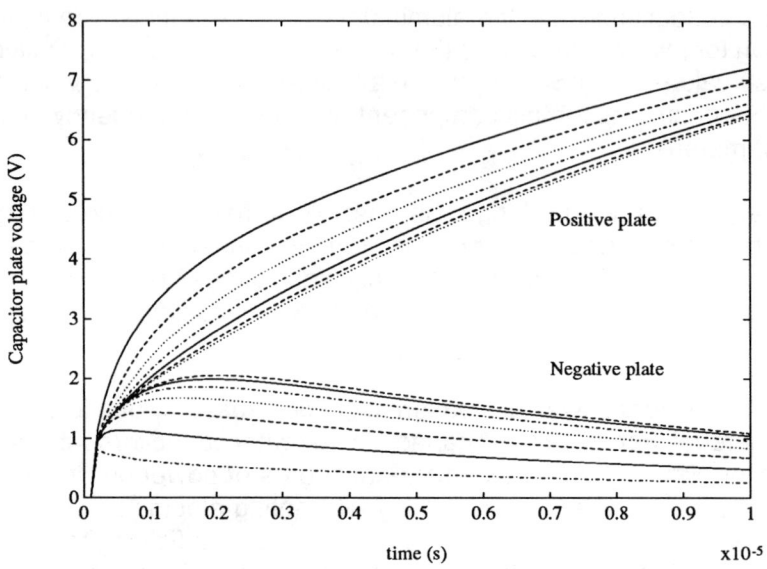

Figure B.18: Local currents

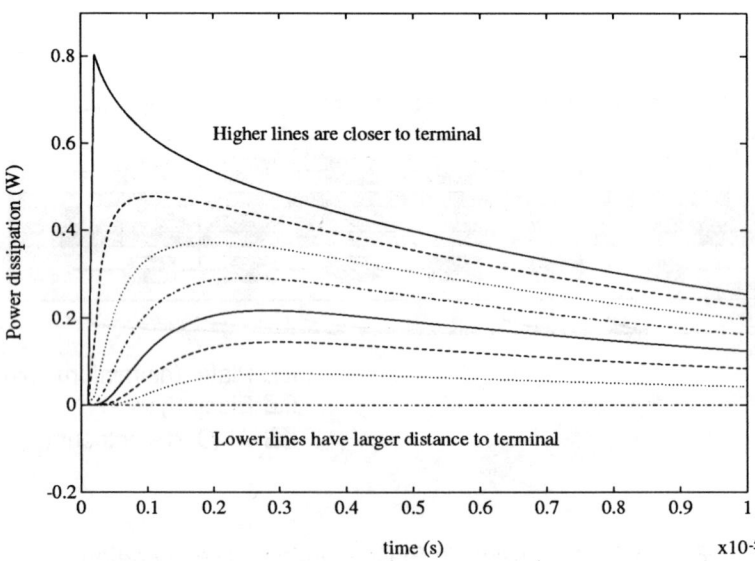

Figure B.19: Local power dissipation

Power dissipation might affect the (often liquid) electrolyte, the dielectric and/or the aluminium foil itself. Summarizing the failure mechanisms for electrolytic capacitors are:

Failure mechanism electrolytic capacitors	Stressor	Related material aspects
High voltage breakdown (Electron trap)	Applied E field pulse-width of E field polarity E field	Amount of impurities in the material used
High voltage breakdown (Impact ionization)	Applied E field, dE/dt polarity E field	Electric breakdown characteristics of the material used
Power overstress	Environmental temperature	
	(Alternating) capacitor current (Pulse power dissipation)	Resistivity capacitor plates
High voltage breakdown (Electron trap)	Applied E field pulse-width of E field	Amount of impurities in the material used
High voltage breakdown (Impact ionization)	Applied E field, dE/dt	Electric breakdown characteristics of the material used

B.3. Inductive components

Most inductive components consist, in principle, of one or more coupled coils. For many inductive devices the core of the device consists of air or other non-magnetic materials. For other inductors the core consists of some kind of magnetic material. Advantage of the use of this magnetic materials in inductive devices it that it improvies the efficiency of the device and concentrates/shields the electromagnetic fields within the device. Relation between voltage and current for a single inductor can be expressed using

$$V = L.\frac{d\,I}{d\,t} \tag{B.3}$$

In practice most inductors will not have this ideal behavior. An often used model for practical inductors is given in figure B.20. Important parameters in this respect are:

— R_s: series resistance of the inductor

— C_p: parallel capacitance of the inductor

— L: inductor (often not constant; depending on the used core material)

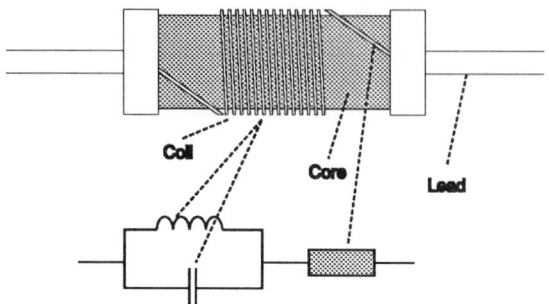

Figure B.20: Inductor model

As the differences between the various inductive devices are considerable the following sections will not discuss general failure mechanisms for inductive devices but will discuss some examples of (simple) inductive devices with their corresponding failure mechanisms.

B.3.1. *Single air coils*

One of the most simple inductive devices is a simple air coil. This device consists of a number of turns of conductive wire, often wound on a frame. This frame consist of non-magnetic material. The principal failure mechanisms of such a device are, due to the similarities in construction, closely related to the failure mechanisms of a wire wound resistor. Main failure mechanism for this class of devices is related to power dissipation due to ohmic losses. An important difference is the possibility of high-voltage breakdown. In a wire wound resistor the windings will be far apart to achieve optimal power distribution. The construction of inductive devices will often show compact winding structures using in many cases multi-layer structures. In those cases where considerable voltages exist between adjacent windings there is the possibility of high-voltage breakdown. Summarizing this gives the following failure mechanisms for single inductors with non-magnetic cores.

Failure mechanism single inductors	Stressor	Related material aspects
Power overstress	Environmental temperature	
	Steady state power dissipation (Hot spot temperature)	Thermal resistance to ambient
	Pulse power dissipation (Hot spot temperature)	Thermal resistance to ambient Thermal capacitance

Failure mechanism single inductors	Stressor	Related material aspects
High voltage breakdown (Impact ionization)	E field between adjacent windings	Electric breakdown characteristics of the isolation material used, construction of the device

B.3.2. *Multiple air coils*

Many inductive devices do not consist of one single coil but of a number of coupled coils. See figure B.14 for a simplified diagram.

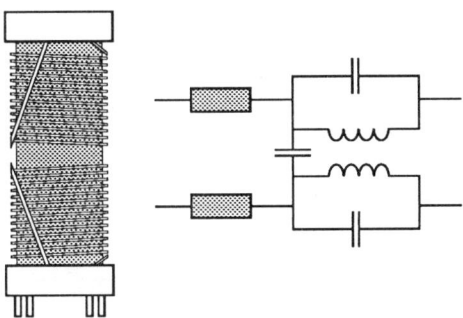

Figure B.21: Model multiple coil

Generally speaking these devices have similar failure mechanisms as single inductors. The most important difference is the possibility of a high- voltage breakdown between two adjacent coils. In many practical situations the E field within a coil will remain within limits but, especially in devices used for isolation purposes, the E field between two coils within a device might be considerably larger. A summary of this gives the following failure mechanisms for inductive devices consisting of multiple coils.

Failure mechanism single inductors	Stressor	Related material aspects
Power overstress	Environmental temperature	
	Steady state power dissipation (Hot spot temperature)	Thermal resistance to ambient
	Pulse power dissipation (Hot spot temperature)	Thermal resistance to ambient Thermal capacitance

Failure mechanism single inductors	Stressor	Related material aspects
High voltage breakdown within coil (Impact ionization)	E field between adjacent windings	Electric breakdown characteristics of the isolation material used, construction of the device
High voltage breakdown between coils (Impact ionization)	E field between adjacent coils	Electric breakdown characteristics of the isolation material used, construction of the device

B.3.3. *Inductive devices using magnetic cores (coils, transformers)*

A disadvantage of the use of inductive devices with non-magnetic core is the relative low efficiency of the device and the spreading of the inductors electromagnetic field. To prevent these problems many inductors use cores of soft-magnetic materials. As a result a major part of the field will be "locked" in the core and, as a second effect, the efficiency of the device will improve. Figure B.22 shows an example of the use of a core in a transformer.

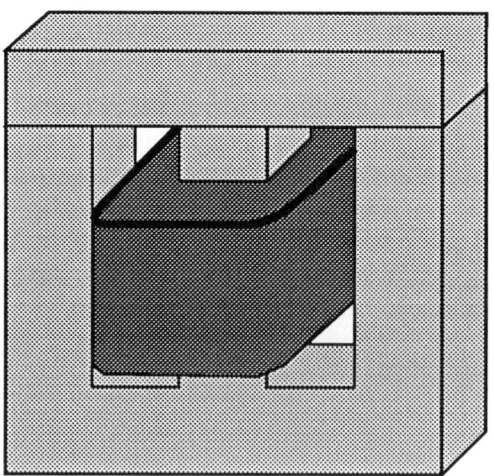

Figure B.22: Transformer

Generally speaking the use of magnetic cores will introduce two effects related to failure probability of a device.

— Hysteresis

— Saturation

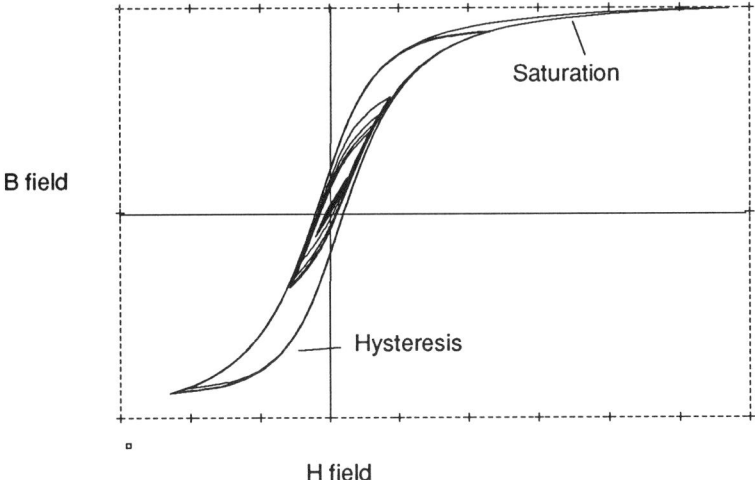

Figure B.23: B-H hysteresis

See figure B.23 for a practical example. For low *B-H* fields the core will show a more or less linear relation between *B* and *H*. Higher *H* fields will show a *B-H* hysteresis. The area within such a hysteresis curve is related to the amount of hysteresis losses. Hysteresis losses will cause power dissipation within the core of the device. At a certain moment increasing H fields will no longer result in increasing *B* fields due to magnetic saturation of the core. Saturation will cause a further increase in power dissipation. Summarizing this gives the following failure mechanisms for inductive devices using cores of (soft) magnetic material.

Failure mechanism single inductors	Stressor	Related material aspects
Power overstress	Environmental temperature	
	Steady state power dissipation (Hot spot temperature)	Thermal resistance to ambient
	Pulse power dissipation (Hot spot temperature)	Thermal resistance to ambient Thermal capacitance
Power overstress due to hysteresis losses	coil currents	*B-H* characteristics used core material Thermal resistance to ambient Thermal capacitance

Failure mechanism single inductors	Stressor	Related material aspects
Power overstress due to saturation losses	coil currents	Saturation B field used core material Thermal resistance to ambient Thermal capacitance
High voltage breakdown within coil (Impact ionization)	E field between adjacent windings	Electric breakdown characteristics of the isolation material used, construction of the device
High voltage breakdown between coils (Impact ionization)	E field between adjacent coils	Electric breakdown characteristics of the isolation material used, construction of the device

B.4. Conclusions

Using elementary failure mechanisms it is possible to obtain susceptibility models for many devices. In this respect it is very important to know what the behaviour of a device will be, not only under normal circumstances, but also under circumstances deviating from normal operation. Side-effects in one component may cause failures in another component. Although the susceptibility models presented in this thesis give by no means a complete overview of all possible failure mechanisms in all components it nevertheless shows the possibilities and guidelines to develop these models. The very important factor 'time' was not taken into account in this chapter. For detailed susceptibility models of practical components it will be necessary not only to develop static susceptibility models but also to look in detail at what happens with susceptibility parameters and functional parameters as a function of time.

Appendix C
Tolerance Models and Examples

C.1. Introduction

As mentioned in Chapter 6 stressor/susceptibility analysis requires, especially for the deriving of realistic stressor sets, the use of detailed component tolerance models. This appendix will present some examples of components where, initially, component tolerance models were unknown. In all the examples the lack of tolerance models hinders not only the stressor/susceptibility analysis but will also cause serious reliability problems in the field.

Due to non-perfect control over the production process of components it is impossible to manufacture series of perfectly identical components. They will all show some deviation from the desired nominal values and a spread in quality.

Components are modelled by sets of parameters (e.g. Spice). An individual component is modeled by choosing certain parameter values. A batch of components can be described by putting tolerances on these parameter values. A tolerance on a certain parameter is described by means of a distribution function, however, knowledge of the distributions of all component parameters does not fully describe a batch of components. In many cases the parameter set chosen to describe a component does not consist of fully independent parameters, so for a complete tolerance model of a component information about parameter correlations is required. Parameter correlations can be described by means of correlation coefficients (see example 1). Once a set of parameters is available to model the functional behaviour of a component (e.g. Spice parameters) two approaches are possible :

— General purpose tolerance model

---

*   Research on component tolerances was carried out by Ir. P.H. Fennema as part of this project.

The distributions of all component parameters are determined by measurements on (large) series of components, selected at random. The correlation coefficients between all possible parameter combinations are calculated. This is a very time consuming and expensive method.

— Application oriented tolerance model

Basically the same as a general purpose tolerance model with one major difference: Only these parameters are taken into account for which the application is very sensitive. For the remaining parameters the nominal value (or an estimation) can be used.

Tolerance models can be used to examine circuit performance variability caused by component tolerances. A well-known example of such an analysis method is the Monte Carlo analysis. Monte Carlo analysis simulates the performance of a batch of circuits. Random values are created for the circuit parameters according to the distribution of each specific parameter. The circuit is analysed by a simulation program. This is repeated a number of times.

The use of knowledge of parameter correlations has certain advantages for these kinds of analysis. When parameter correlations are not taken into account values will be given to every individual parameter according to its own specific distribution function. In practice, however, some combinations of parameters, which can occur when this strategy is used, will never exist at all. For example when two parameters have a high correlation and the first one has a value at the upper tail of its distribution, the second one will also have a value at the upper tail of its distribution, and not at the lower tail (which easily could have been the case when parameter correlations had not been taken into account). When taking parameter combinations into account, only realistic combinations of parameter values are generated, so the accuracy of the batch performance analysis increases. An extra advantage is that the number of runs for a Monte Carlo analysis can be reduced.

C.2. **Example 1: Foil transformer (circuit B)**

The transformer used in circuit B is a foil transformer. The complete model of this transformer consists of 2 Q-parameters, 10 L-parameters, 10 R-parameters (representing the series resistances of the windings) and 7 C-parameters (representing the capacitances between foil layers). A total series of 250 transformers, selected at random, was examined. All transformer parameters have been measured. An example of the measured distribution of one of the parameters is shown in figure C.1.

Figure C.2 describes the correlations between parameters. It is a three dimensional plot of a correlation matrix C. Element $C[i,j]$ describes the correlation coefficient between the two parameters having indices *i* and *j* respectively. For a correlation matrix always $C[i,j]=C[j,i]$ and $C[i,j]=1$ for $i=j$.

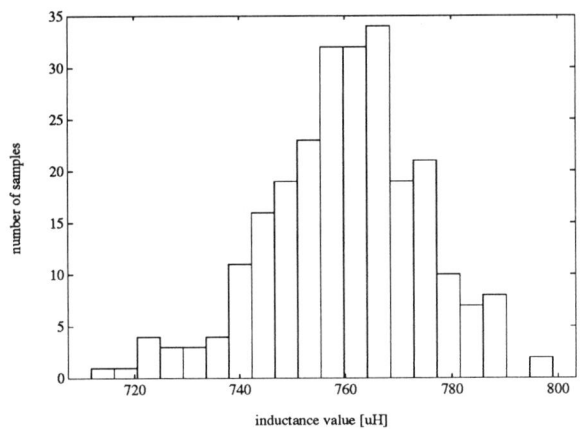

Figure C.1:Parameter distribution primary inductances

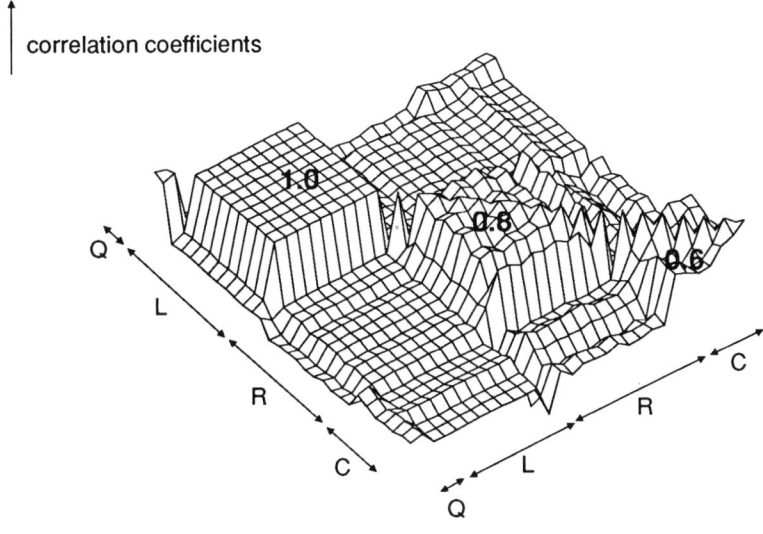

Figure C.2:Correlation diagram transformer parameters

The correlation plot shows that especially *L* parameters have a high correlation (the large plateau of altitude 1.0 in figure C.2). This information can be very useful when generating pseudo-random values of transformer parameters in a Monte Carlo analysis.

A very important aspect of developing tolerance models is the way the parameter values are measured. This is illustrated by the following example. The Q parameters of the transformer were originally specified at a measurement frequency of 1 kHz. In the application however, the frequency range is between 10 kHz and 100 kHz. The results of the examination of the frequency dependency of the Q factor are shown in figure C.3. The picture shows that in this case a measurement at 1 kHz is not useful as a specification. At 1 kHz it is hard to distinguish between transformers within specifications and transformers outside specifications. For this reason the specification was changed to a frequency of 10 kHz (not higher due to practical limitations). Whenever tolerance models are developed the purpose of the model should always be considered (general purpose or application oriented).

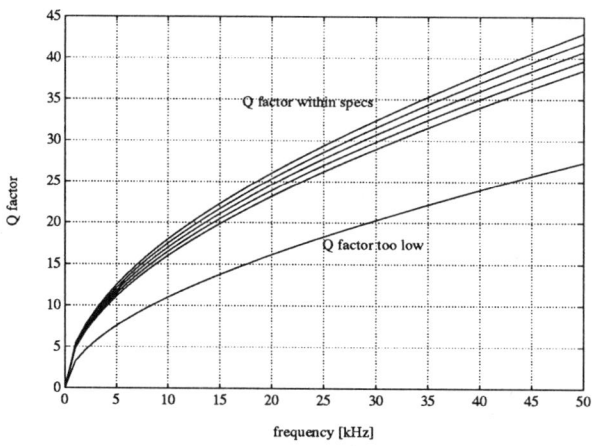

Figure C.3:Frequency dependency of the Q factor

C.3. Example 2: The optocoupler (circuit B)

An optocoupler consists of a light emitting diode (LED) and a phototransistor (figure C.4). The base collector junction of the transistor is the light collecting element, represented as a diode. The photocurrent I_{ph} is generated in this diode and amplified by a factor β_f (the transistors current gain). The reason for the development of a tolerance model for this optocoupler is directly related to the application. In practice the switching speed of the optocoupler appears to be a very important parameter. An optocoupler which switches on too slow can cause instable behaviour of the power supply in stand by

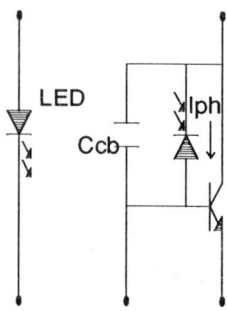

Figure C.4:Optocoupler

mode. Because the power supply appears to be very sensitive for the switching speed of the optocoupler, the development of a tolerance model for the opto-coupler is mainly concentrated on parameters related to the switching speed.

First of all a series of optocouplers was measured. The series consisted for a part of optocouplers selected at random and for a part of optocouplers obtained from failing power supplies. The following parameters were measured in a test circuit described in the optocoupler databook :

— t_{on} (on time, see [Schm75] for an exact definition)

— t_f (fall time, see [Schm75] for an exact definition)

— t_{onps} (on time in power supply)

— CTR (current transfer ratio)

— β_f (current gain phototransistor)

Results (see also figures C.5 to C.7):

— Optocouplers which switch on slow (large t_{on}) in the test circuit also switch on slow in the power supply (large t_{onps}).

— Optocouplers having a high CTR have a low t_{on} (switch on fast).

— A high correlation exists between the current amplification β_f and the fall time t_f.

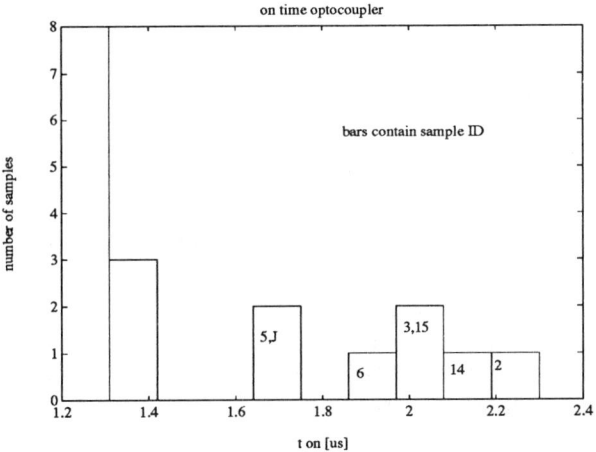

Figure C.5:t_{on} (on time optocoupler)

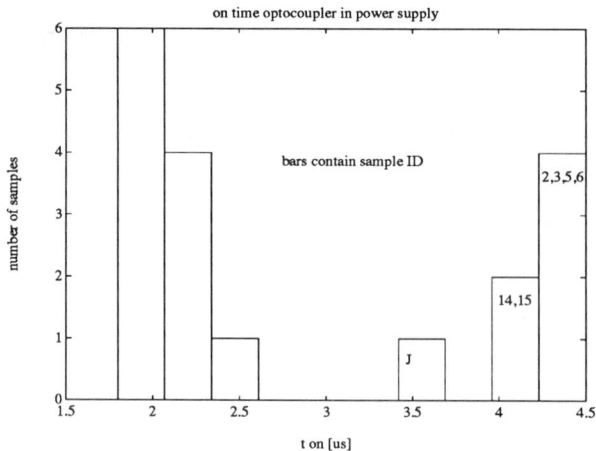

Figure C.6:t$_{onps}$ (on time optocoupler in power supply)

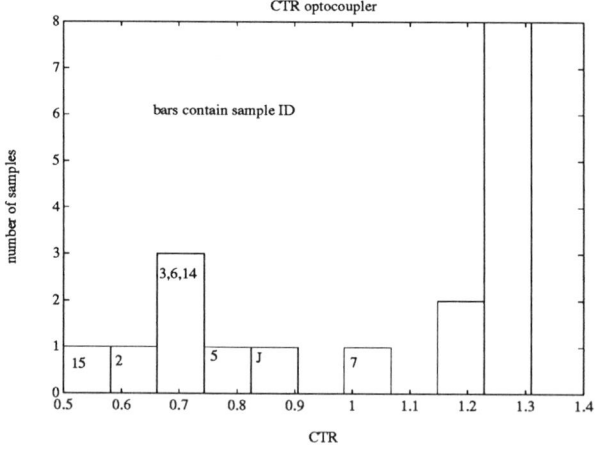

Figure C.7:CTR optocoupler

As a first and fast solution, practical problems with failing power supplies could be decreased by only using optocouplers with a high CTR. This high CTR selection reduces the probability of using slow optocouplers.

For development of a tolerance model it is necessary to go into more detail and to examine the physical backgrounds of the switching behaviour of the optocoupler. Which parameters influence the switching speed and what is the physical meaning of the parameter correlations found in the measurements?

The switching speed of an optocoupler is mainly determined by the phototransistor [Schm75] . A phototransistor has a relatively large collector base surface (in order to obtain a sufficient large photoelectric sensitivity). This results in a large collector base capacitance C_{cb}, which is the basic cause of the relatively slow switching behaviour.

The effect of the capacitance C_{cb} on the switching behaviour of the optocoupler is, however, strongly influenced by other parameters:

— optoelectrical parameters : an optimal transfer of energy from the diode, through the optical medium to an electrical photocurrent in the base of the transistor is very important. The more efficient this transfer, the higher the currents through the transistor. C_{cb} will be loaded faster, resulting in a lower switch on time. Optoelectrical parameters do not influence the switching off behaviour.

— current gain of the phototransistor (β_f) : By the Miller effect the collector base capacitance C_{cb} appears as a capacitance of β_f times C_{cb} at the output of the phototransistor. So the effective capacitance causing the slow switching is in fact a factor β_f larger than C_{cb}.

By examining the measurements results it is possible to determine the influence of the optoelectrical parameters and the current gain on the switching times. First it is necessary to examine the CTR in more detail:

The CTR is determined by the optoelectrical parameters and by β_f.

— optoelectrical parameters : The more efficient the energy transfer mentioned above, the higher the CTR.

— current gain of the phototransistor (β_f) : The higher β_f, the higher the CTR.

Main influence factors for switch on time (t_{on}) :

The CTR is influenced by both the photoelectrical parameters and β_f. Because in practice a high CTR corresponds with a low t_{on}, apparently ton is mainly influenced by photoelectric parameters. If the influence of β_f would be dominant a high CTR would correspond with a high ton.

Main influence factors for the fall time (t_f) :

The fall time is not influenced by the photoelectrical parameters. It is determined by C_{cb} and β_f (Miller effect). Results of the experiments show a high correlation between Bf and the fall-time.

C.4. **Example 3 : High voltage transistor**

Requirement for the development of a tolerance model is that parameters are adapted to the simulation tool used (Spice). The problem is how to obtain the distributions of the Spice parameters. Two different approaches are possible.

— Spice parameters are not independent. They are all based upon the physical properties of the transistor. These properties are in fact determined by the production process of the transistor. Tolerances in Spice parameters are determined by tolerances in production parameters. So it is possible to express the Spice parameters in terms of production process parameters. By investigating the production process and tolerances in process parameters it is possible to obtain the Spice parameter distributions.

— Another possibility is to measure the Spice parameters of a series of transistors (see figures C.8 and C.9). Parameter extraction methods can be found in [Get78]. From the results correlations between parameters can be calculated.

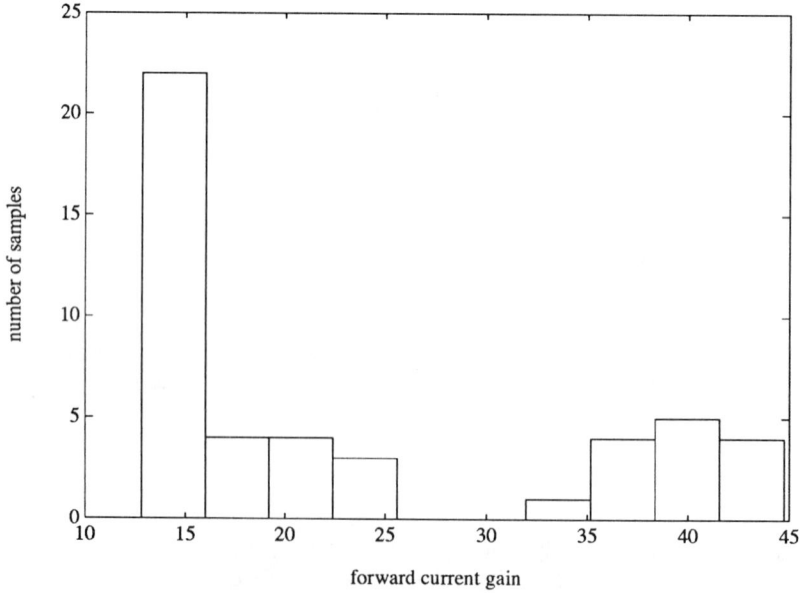

Figure C.8:Parameter distribution forward current gain

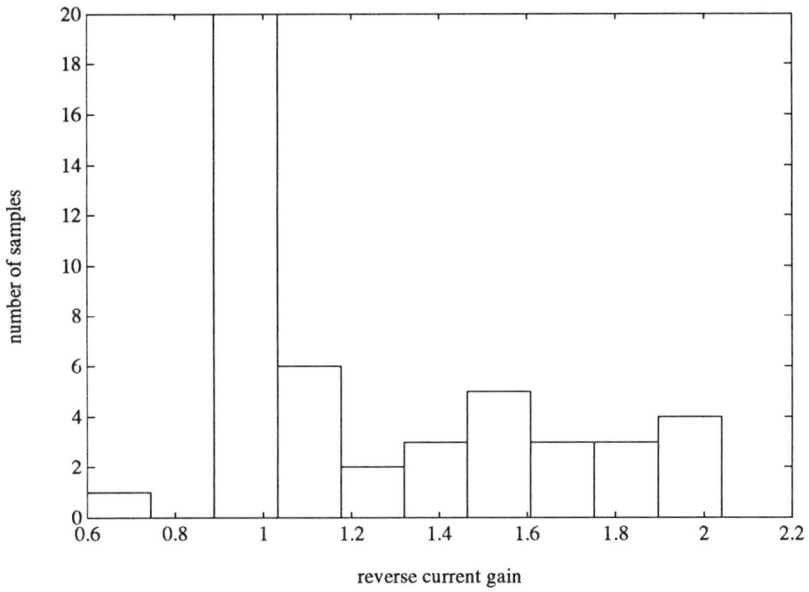

Figure C.9:Parameter distribution reverse current gain

C.5. **Conclusions**

The previous sections showed some examples of component tolerance analysis and the resulting component tolerance models developed during this project. As the main purpose of this project was to obtain of reliability models and methods, component tolerance models were only developed on the level of application oriented tolerance models. The presented models are by no means complete component tolerance models. It is, however, possible to derive demands from this chapter for the development of component tolerance models required for use in either functional tolerance analysis and/or stressor susceptibility analysis. A component tolerance model should cover the following effects.

— A component tolerance model should present component parameters, important for the functional behaviour of the component, each with their related distribution.

— A component tolerance model should present correlation factors indicating the correlations between the various component parameters.

Analysis of the practical circuits, mentioned in this book, showed that for most components tolerance models, even on the level of application oriented models, were totally unavailable. This lack of models proved to be a major problem, not only during stressor/susceptibility analysis or functional tolerance analysis but during the entire analysis of these circuits.

Appendix D
Bibliography

Ame86   Amerasekera E.A., Campbell D.S., *Oxide breakdown in MOS structures under ESD*, Proceedings RELCON'86 conference on Reliability Technology, Lynby, Denmark, 1986

Ame87   Amerasekera E.A., Campbell D.S., *Failure mechanisms in semiconductor devices*, John Wiley & Sons, 1987

Bas86   Basso L., Winterbottom A., Wynn H.P., *A Review of the 'Taguchi Methods' for Off-line Quality control*, Quality and Reliability Engineering International, 1986

Bot85   Boettger H. Bryksin V.V., *Hopping conduction in Solids*, VCH Publishers, Deerfield Beach, (Fla), USA, 1985

Bro88   Brombacher, A.C., *Interim report cooperation Philips Videq - Twente University; Reliability prediction self oscillating power supply*, Twente University of Technology rep. nr. 88D031, 1988

Bro89a   Brombacher A.C., de Boer H.A., van't Loo J., *Integration of Reliability & Tolerance Effect Analysis*, Proceedings IEEE Annual Reliability, Availability and Maintainability Symposium, Atlanta (GA), USA, 1989

Bro89b   Brombacher A.C., Fennema P.H., van't Loo J. *Combined Reliability and Tolerance Analysis with Emphasis on Failure Causes of Key-Components*, VDI Symposium on Reliability of Components in Technical Systems, Munich, Germany, 1989 (in German)

Bro89c   Brombacher A.C., Vos R., Fennema P.H., van't Loo J., *The robust design of electronic systems*, ASQC Quality in Electronics conference, San Jose (Ca), USA, 1989

BT87   *Handbook of Reliability Data for Components used in Telecommunications Systems*, Issue 4, British Telecom, 1987

Eli75   Elias N.J., *New statistical methods for assigning device tolerances*, Proceedings IEEE International Symposium on Circuits and Systems, Newton, Mass, 1975

Get78   Getreu I.,*Modeling the bipolar transistor*, Elsevier, 1978

Hum88   Humphreys M.J., *A study of the Failure Locus of NPN transistors and its Improvement using Graded Collector Structures*, PhD thesis University of Manchester UK, 1988

Jen89   Jensen F.,*Component Failures Based on Flaw Distributions*, IEEE Annual Reliability, Availability and Maintainability Symposium, Atlanta (GA), USA, 1989

Jon89   Johnson B.W., *Design and Analysis of Fault-tolerant Systems for Industrial Applications*, 4th International GI/ITG/GMA symposium on Fault-Tolerant Computing Systems, Baden-Baden, Springer verlag 1989

Kac85   Kackar R.N., *Off-line Quality control, Parameter Design and the Taguchi Method*, Journal of Quality Technology, October 1985

Klo89   Kloosterman W.J., de Graaf H.C., *Avalanche Multiplication in a Compact Transistor Models for Circuit Simulation*, IEEE transactions on electron devices, July 1989

Lee88   Van Leeuwen P.G., *Application of an EMF sensing scanner drive IC*. Philips internal report nr. AR82.2- RO78/88, 1988 (Available on request from Philips M.R. laboratory, Eindhoven, the Netherlands)

Let84   Leturcq P., *Power Bipolar Devices*, Microelectronics and Reliability, vol.24, no2, pp.313-337, 1984

Leu74   Leung K.H., Spence R., *Efficient statistical circuit analysis*, Electronic letters 10, pages 360-362, 1974

Mar84   Marinus, A. Self Oscillating Power Supplies for system 5, Philips internal report nr. AR6-1736E, Philips TV Europe, Eindhoven, the Netherlands, 1984

MIL65	*Military standardization handbook Reliability Prediction of Electronic Equipment MIL- HDBK-217A*, United States Department of Defense, 1965
MIL87	*Military handbook Reliability Prediction of Electronic Equipment (MIL- HDBK-217E)*, United States Department of Defense, 1987
Mul86	Muller R.S., Kamins T.I., *Device Electronics for Integrated Circuits*, John Wiley & Sons, 1986
Mul86a	Muller R.S., Kamins T.I., *Device Electronics for Integrated Circuits*, John Wiley & Sons 1986, page 202
Mul86b	Muller R.S., Kamins T.I., *Device Electronics for Integrated Circuits*, John Wiley & Sons 1986, page 248
Pan84	Pantic D.M., *Maturity Factors in Predicting Failure Rate for Linear Integrated Circuits*, IEEE transactions on Reliability, nr. 33-3, August 1984
Pec88	Pecht M., Kang Wen-Chang, *A Critique of MIL-HDBK-217E Reliability prediction methods*, IEEE transactions on Reliability Nr. 37-5, December 1988
Phi88	*Reliability Prediction Failure Rates*, Philips internal standard nr. UAT-0387, Philips Eindhoven, 1988
RCA70	*Power Transistors*, for Amplification, Switching, and Control. RCA Technical Series PM-80, Sommerville, NJ, 1970
Rey82	Reynolds F.H., *Measuring and Modeling Integrated Circuit Failure Rates.* British Telecom Research Laboratories, Ipswich, UK, 1982
Sch86	Schaik G. v., Ulenaers M., *Orientation investigation on 3A Schottky rectifiers of Motorola and GI*, Philips CE Component Investigation and Reliability Group (CIRG), Philips internal report number DA635276, 1986
Schm75	Schmidt W.,*Optoelektronik kurz und bundig* (in German), Vogel-Verlag (Germany), 1975
Sho68	Shooman, M.L., *Probablistic Reliability, an Engineering Approach*, Mc Graw Hill, New York, 1968

Sie82   Siewiorek, D.P., Swarz, R.S., *The Theory and Practice of Reliable System Design*, Digital press, Digital Press, Bedford (Mass.), 1982

Spe88   Spence R., Singh Soin R., *Tolerance Design of Electronic Circuits*, Addison Wesley, 1988

Spe88a   Spence R., Singh Soin R., *Tolerance Design of Electronic Circuits*, Addison Wesley 1988, pages 30-32

Spe88b   Spence R., Singh Soin R., *Tolerance Design of Electronic Circuits*, Addison Wesley 1988, pages 126-129

Spe88c   Spence R., Singh Soin R., *Tolerance Design of Electronic Circuits*, Addison Wesley 1988, page 129

Tod70   Todd, C.D., *Zener and Avalanche Diodes*, Wiley-Interscience 1970

Wea87   Weast R.C., Astle M.J. (Editors), *Handbook of Physics and Chemistry*, CRC press, 1987

Tri88   Trip A., *Taguchi methoden voor het ontwerpen van Kwaliteit*, (In dutch) Internal publication of the Philips Center for Quantitative Methods (CQM) Nr. 72., Philips Eindhoven, 1988

Wad85   Wada T., Higuchi H., Ajiki T., *Electromigration in Double- Layer Metalization*, IEEE Transactions on Reliability, April 1985

Wal89   Walinga J.S., *Development of a Carad module for failure rate prediction and load sensitivity analysis*, Twente University internal report nr. EL/BSC/89N102, 1989

Wou86   Wouw T. vd., *Tutorial High Voltage Switching Devices*, Philips Components Hazelgrove internal publication, 1986

Wou86a   Wouw T. vd., *Tutorial High Voltage Switching Devices*, Philips Components Hazelgrove internal publication. 1986, page 17

Appendix E
Terms Used

E.1. General terms

part	Electronic element, defined using "black box" approach. It is not usual to define sub-levels within a part.
component	Electronic device used as functional element.
circuit	Electronic structure of components, intended to perform a certain task.
system	Electronic structure, consisting of one or more (sub-) systems or circuits.
batch	A (large) number of systems, circuits or components with identical structure.
failure feedback	The process of collecting components, circuits and systems, failed during actual use, for analysis on failure causes.

E.2. Stressor / susceptibility

stressor ψ	A stressor is a physical entity influencing the lifetime of a component or circuit. A stressor, indicating an entity X is denoted as ψ_x.
basic stressor	A basic stressor is defined as a stressor with direct, physical influence on a failure mechanism of a component. It is only possible to describe basic stressors partly in terms of other basic stressors.
stochastic stressor $\underline{\psi}_x$	A stochastic stressor, denoted as $\underline{\psi}_x$ is a random function, where X characterizes the type of stressor under consideration.

Note: Unless mentioned otherwise the term stressor is used to describe stochastic basic stressors

stressor set   The collection of basic stressors influencing one failure mechanism.

individual   The ergodic stressor set of one single circuit covering all states
stressor set   and transitions between states of this circuit.

mean stressor   The average of a series of individual stressor sets representing
set   the stressors in a batch of circuits.

susceptibility The susceptibility of a component to a certain failure mechanism
S_Y is defined as the probability that a component will not remain
                operational for a certain period of time under a given combination of stressors. The susceptibility to a failure mechanism Y is
                denoted as S_Y.

mean   The average susceptibility of a batch of components to a certain
susceptibility failure mechanism.

E.3. Traditional reliability prediction models

reliability, *R(t)*   The probability *R(t)* that a component, circuit or system is able
                       to perform all its intended functions during a certain period of
                       time.

failure proba- The probability *F(t)* that a component, circuit or system is not
bility, *F(t)* able to perform one or more of it's intended functions during a
                certain period of time.

failure rate, λ Rate of occurrence of failures in failures/hour or fit (failures/million hour).

λ_p Part failure rate (failures / hour). Failure rate of a component
                taking into account all applicable correction- and acceleration-
                factors.

λ_b Part basic failure rate (failures / hour). Failure rate of a component *not* taking into account correction- and acceleration- factors.

π_x Acceleration- or correction- factor for effect x. Well-known π
                factors are the environmental acceleration factor π_e and the
                quality correction factor π_q.

T_j Junction temperature in K or ^{o} C.

E_a Activation energy of a (degradation) process (eV).

$A, B, T_m, T_r,$ Failure rate scaling or shaping parameters. These parameters
N_t, P, φ are used in the calculation of the λ_b of a component.

$C_1, C_2, C_3,$ Complexity factors for integrated circuits depending on the
N_g, G number of gates/transistors, number of pins, etc.

k Boltzmann constant ($8.63 \cdot 10^{-5}$ eV/k).

PCA Part Count Analysis. A method to predict the failure rate of
components using fixed, estimated, stress factors.

PSA Part Stress Analysis. A method to predict the failure rate of
components using stress factors derived from the (mean)
stresses in the application.

E.4. Electrical variables

V Terminal or node voltage (V).

I Terminal or branch current (A).

ρ Resistivity Ωm.

Q Charge (C)

Q_s Stored charge (C) (especially in semiconductors).

N Concentration charge carriers.

N_+ Concentration holes.

N_- Concentration electrons.

E Electric field (V/m).

Index

A

acceleration factor *11, 16, 18, 20, 23*

acceleration tests *11*

acquisition of stressor sets *136, 156*

activation energy *201*

actual measurements *139*

ageing, definition of *58*

application class *208*

application oriented tolerance model *254*

Arrhenius Law *10, 18, 29, 201*

avalanche and zener *51*

avalanche breakdown *94, 117, 119, 121, 126, 131*

average stress *35*

B

basic stressors, definition of *41*

batch optimization *175*

bottom- up method *40*

British Telecom handbook HRD-4 *11, 18, 24, 214*

C

capacitive components, susceptibility model *241*

catastrophic failure *182*

centre of gravity method *185*

centre of gravity method, enhancements *190*

ceramic/plastic capacitors *243*

charge bubble *114*

circuit simulation *139, 178*

CNET, failure rate model for integrated circuits *21*

coils, transformers *250*

combined interaction effects *77*

component models *141*

component screening techniques *208*

component tolerance model *261*

component tolerance models *145*

component tolerance models *253*

computer simulation *139, 204*

constantly degrading susceptibility models *58*

consumer electronics *208*

correlation *146, 253*

correlation factors *261*

correlation plot *255*

corrosion *51, 96, 37, 115, 137*

current breakdown *51, 89, 103, 115, 117, 121, 126, 130, 137*

current crowding *106*

D

databooks *145, 203*

degradation effects *75*

degrading susceptibility models *58*

derating *32*

deriving stressor sets (practical measurements) *148*

design parameters *77*

design process *174*

designable parameters *176, 201*

deterministic exploration *180*

device structure *208*

dielectric material *244*

discrete probability density function *44*

discrete probability function *44*

distribution function *254*

dynamic systems *40*

E

Ebers-Moll *143*

effective device temperature *208*

electrical overstress failure mechanisms *85*

electrical stressors *51*

electro-mechanical *229*

electrolytic capacitors *244*

electromigration *51, 96, 98, 115, 126, 130 137*

electron trap ionization *51, 95*

ergodicity *135*

exploration points *179*

extended MIL-HDBK-217E model, integrated circuits (Pantic model) *22*

F

failure mechanisms, definition of *40*

failure mechanisms, simple components *233*

failure mechanisms *202*

failure prediction *39*

failure probability and reliability *62*

failure probability, multiple failure mechanisms *65*

failure probability, single failure mechanisms *62*

failure probability, definition of *12*

failure rate, definition of *13*

failures obtained using reliability tests *10*

feedback *153*

field data *10, 24, 208*

foil transformer *254*

forward bias second breakdown *52, 110, 116, 121, 123, 127, 138*

functional component models *143*

functional tolerance design *204*

G

Gaussian distribution *180*

general failure mechanisms, capacitors *242*

general failure mechanisms, resistors *233*

general purpose tolerance model *253*

geometrical transistor aspects, breakdown effects *106*

gradual failure mechanisms, cumulative effects *76*

gradual loss functions *190*

gradual susceptibility models *56*

Gummel-Poon *143*

H

hazard rate *70*

high voltage transistor *121, 137*

high-voltage switching transistor *84*

high-voltage breakdown *51, 93, 115, 137*

high-voltage breakdown, capacitors *242*

high-voltage transistor *156*

hot-spot *110*

hot-spot melting *89*

hysteresis *250*

I

impact ionization *51, 93*

impurities *242*

inductive components *247*

inherently robust *191*

iso-probability density lines *160*

J

joint parameter distribution diagram *180*

joint probability *180*

K

Kasouf and Mercurio model (integrated circuits) *20*

L

long-term failure mechanisms *96*

long-term stressor models *189*

long-term susceptibility models *189*

M

manufacturer failures *169*

Markov approach *66*

mean stressor probability density function *47*

measurement of individual stressor-sets *150*

measuring stressors *204*

mechanical properties *229*

mechanical stressors *53*

MIL-HDBK-217 *11, 15, 24, 214*

MIL-HDBK-217 motor model *229*

minimal sensitivity *192*

minimal variability *192*

model parameters *144*

Monte Carlo *177*

Monte Carlo analysis *181, 254*

motor drive circuit *84, 124, 208*

multi variable catastrophic susceptibility models *55*

multi variable stressor probability density functions *49*

multi parameter devices *204*

multiple air coils *249*

mutual influence between parts *201*

N

nominal design *174*

O

one variable catastrophic susceptibility model *54*

optimization (entire batch) *175*

optimization (nominal circuit) *175*

optimum reliability *197*

optocoupler *173, 256*

P

parameter optimization *177*

parameter regionalization *179*

parameter tolerances *196*

Part Stress Analysis *15*

Part Stress Count analysis *18*

Parts Count Analysis *15*

pass-fail diagrams *177, 182*

Philips failure rate model (ICs) *20*

Philpac *204*

phototransistor *256*

pinch-in *106, 111, 112*

post-mortem analysis *153*

power breakdown *91, 103, 117, 120, 121*

power breakdown, capacitors *243*

power breakdown, resistors *233*

power dissipation *130*

power dissipation, electrolytic capacitor *245*

power interrupts *174*

practical circuits *135, 207*

practical reliability figures *24*

practical stressor/susceptibility interactions *156*

probability of omission *179*

problem of dimensionality *181*

production process *253*

pseudo- Gaussian distribution *180*

pulse power effects *100, 117, 118, 132*

pulse power effects, resistors *234*

Q

quasi- stationary states *149*

R

random failures, definition of *13*

random time *174*

rating *32*

reliability calculations (standard reliability prediction handbooks) *24*

reliability figures *207*

reliability optimization *29, 175, 176, 204*

reliability optimization (computer aided) *177*

reliability prediction *39*

reliability prediction handbooks *207*

reliability prediction methods *9*

reliability prediction models *11*

reliability, definition of *12*

resistive components *233*

reverse bias second breakdown *52, 111, 116, 121, 123, 127, 138*

robust *191*

S

safe operating area (SOA/SOAR) *103, 175*

saturation *250*

schottky diode *84, 117, 137, 156*

second breakdown *103, 105, 133*

secondary diffusion *51, 96, 99, 115, 121, 130, 138*

self oscillating power supply *208*

sensitivity *191*

simulation software *141*

SPICE *143, 204*

spread (quality) *253*

square planar transistor *108*

standard environment class *208*

statistical exploration *181*

stress *35*

stressor measurement *150, 168*

stressor probability density function *42*

stressor sets *204*

stressor, definition of *40*

stressor / susceptibity interaction *39*

stressor / susceptibility diagram *175*

stressors, stochastic function *42*

susceptibility *36*

susceptibility density function, definition of *54*

susceptibility limit *182*

susceptibility models *203*

susceptibility models for practical components *117*

susceptibility models, large series components *59*

susceptibility, definition of *53*

switch-off effect *52, 138*

switch-off effect (diodes) *116*

switch-off power loss *126*

switch-on effect *52, 138*

switch-on effect (diodes) *115*

switch-on power loss *126*

T

Taguchi *190*

test circuits *207*

thermal considerations *85*

thermal cracks *51, 91, 115, 137*

thermal stressors *53*

tolerance effects *175*

tolerance models *141, 194*

tolerance space *178*

top-down method *39*

traditional reliability analysis/prediction methods *207*

V

variability *192*

variability] *254*

vertex analysis *178*

video cassette recorders *207*

voltage breakdown *103*

W

weak sub-populations *61, 74*

Weibull *230*

wire wound resistors *238*

worst-case analysis *178*

Z

Zener breakdown *94*